Anthony

**Structure
and function
of the
body**

*Third
edition*

Structure and function of the body

Catherine Parker Anthony
R.N., B.A., M.S.

Formerly Assistant Professor of Nursing, Science Department,
Frances Payne Bolton School of Nursing, Western Reserve University;
formerly Instructor of Anatomy and Physiology,
Lutheran Hospital, Cleveland, Ohio; formerly instructor
of Anatomy and Physiology, St. Luke's Hospital, Cleveland, Ohio;
formerly Assistant Instructor of Anatomy and Physiology,
Frances Payne Bolton School of Nursing,
Western Reserve University

With 76 illustrations

Third edition

The C. V. Mosby Company

Saint Louis   1968

Preface

The purpose of this book is to help students learn and teachers teach basic information about the human body effectively, efficiently, and enjoyably. Organization of the book's contents contributes to the fulfillment of these aims by making facts and principles stand out clearly. For example, topical outlines begin each chapter so students and instructors can quickly preview its contents. Summarizing outlines and review questions close each chapter. These aids can guide students in their learning and review of information. They can also help instructors plan classroom discussions and examinations. Illustrations, uncluttered by excessive details, and summarizing tables are other features of this book which facilitate learning and teaching.

Concrete examples that tie facts and principles to practical situations abound in this third edition of *Structure and Function of the Body*. Included are many correlations between normal and abnormal structure and function. Normal structure and function are also related to various laboratory tests used to help diagnose disease and to various medical and nursing procedures used to treat disease. Such correlations, I hope, breathe life into facts and make for more effective learning as well as easier and more enjoyable reading.

Those readers familiar with the previous editions of this book will probably like to know about specific changes made in this edition. Chapter 1, for example, contains considerably more information about cell structure and function; it also now includes a discussion of the pleura and the peritoneum. Information about tendons and bursae has been added to the chapter on muscles. The chapter on the nervous system has been reorganized; some of it has been revised and a considerable amount of new material has been added to it. Other subjects added or revised include the following: blood types, coronary arteries, blood pressure, chemical digestion, and hormonal control of metabolism, of kidney functions, and of stress responses. New ideas about the thymus gland and immunity have also been included.

I express warm appreciation to the artist, Mrs. Mary Farr, to my secretary, Mrs. Georgeanna Keefe, and to hundreds of teachers and students whose ideas, talents, and stimulation have helped make this book.

Catherine Parker Anthony
Cleveland, Ohio

Contents

Introduction

I

General plan and structural units of the body

Have you ever asked yourself any questions like these—Why do I have to breathe? Why do I have to eat? Why do I have to have water? What does my body do with air, food, and water? Why do I seem to need more air, food, or water sometimes and less other times? How does my body keep itself warm? Even more perplexing, how does it keep itself almost as warm when I am out in a raging blizzard as when I am under the blistering sun? How do I move, and in so many different ways? Why is it that such different diseases as poliomyelitis, strokes, and arthritis can all make me unable to move normally? How do I see? Hear? Think? Feel?

In this book you will find the answers to these and many other questions. Chapter 1 describes the body's general plan, its main structural parts, and its chief functions.

GENERAL PLAN

One way that you can think of the body is as a building with only four rooms. The "rooms" are called cavities. The four cavities of the body are: the chest cavity, the abdominopelvic cavity, the cranial cavity, and the spinal cavity. ("Thoracic cavity" is the more technical anatomic name for the chest cavity, which is the space enclosed by the ribs. The abdominopelvic cavity is often referred to as the abdominal and pelvic cavities, as if it were two cavities instead of one. Actually, however, it is only one cavity because no partition of any kind divides it into two.) The *cranial cavity* is the space inside the skull, and the *spinal cavity* is the long cylindrical space inside the spinal column.

Look at Figure 1-1 to locate the body cavities just named. Note the partition shown between the thoracic and abdominal

3

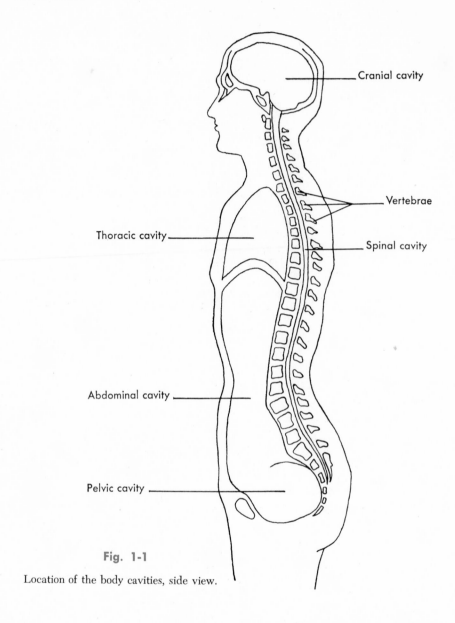

Cranial cavity

Vertebrae

Spinal cavity

Thoracic cavity

Abdominal cavity

Pelvic cavity

Fig. 1-1

Location of the body cavities, side view.

cavities. This represents the diaphragm muscle.

Figure 1-2 shows some of the organs contained in the largest body cavities. In the thoracic cavity, for example, are located the lungs and heart, plus the aorta and some other vessels, various nerves, and the thymus gland. The abdominal cavity contains the stomach, liver, gallbladder, small intestine, most of the large intestine (colon and cecum), pancreas, appendix, and spleen. The two kidneys lie behind the abdominal cavity, right underneath its lining. The reproductive organs, urinary bladder, and lowest part of the intestine fill the pelvic cavity. The brain occupies the cranial cavity, and the spinal cord occupies the spinal cavity. Find each body cavity in a model of the human body if you have access to one. Try to identify the organs in

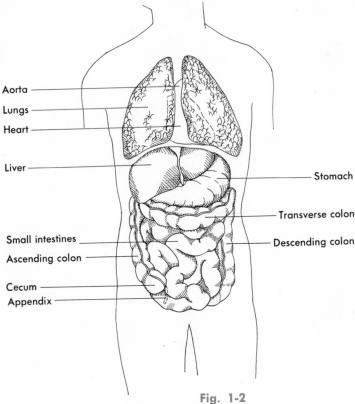

Aorta
Lungs
Heart
Liver
Stomach
Transverse colon
Small intestines
Descending colon
Ascending colon
Cecum
Appendix

Fig. 1-2

Organs of the chest and the abdominal cavities, viewed from the front.

each cavity. Try to visualize their locations in your own body. Study Figures 1-1 and 1-2.

A moist smooth slippery membrane—the *pleura*—lines the thoracic cavity and forms a thin transparent covering on the surface of each lung. The term *parietal pleura* means the lining of the thoracic cavity while *visceral pleura* designates the covering of the lung. Between the parietal pleura and visceral pleura lies the small *pleural space*. Normally it contains just enough fluid to make both portions of the pleura moist and slippery and able to glide easily against each other as the lungs expand and deflate with each breath. Abnormally, the pleural space sometimes fills up with a large amount of fluid. The extra fluid presses on the lungs and makes it hard for a patient to breathe, so that a physician may decide to remove the excess pleural

fluid. To do this he thrusts a hollow tube-like instrument through the patient's chest wall into his pleural space, and the excess fluid drains out. If you were to ask this patient what the doctor did to him, he would most likely answer, "He tapped my chest." If you asked the doctor he would probably answer with the technical name of the procedure. He would say that he had done a *thoracentesis* on the patient.

Pleurisy and pneumothorax are other terms that have to do with the pleura and pleural space. *Pleurisy* (or pleuritis) is an inflammation of the pleura. *Pneumothorax* is the injection of air into the pleural space. The additional air presses on one lung and collapses it. It stays collapsed for some time—in other words, is not used for breathing. Pneumothorax is therefore a procedure performed to rest a diseased lung.

A membrane similar in structure and function to the pleura lines the abdominal cavity and covers the surfaces of abdominal organs. It is called the *peritoneum. Abdominal paracentesis* is the technical name for the procedure laymen call tapping the abdomen. It consists of introducing a hollow instrument into the peritoneal cavity in or-der to drain off excess peritoneal fluid. How would you define the term "peritonitis"?

STRUCTURAL UNITS

Just as any building is made up of many kinds of structural units (walls, floors, steel, glass, bricks, nails, for instance), so too the body is made up of different kinds of structural units. The body's structural units are called cells, tissues, organs, and systems.

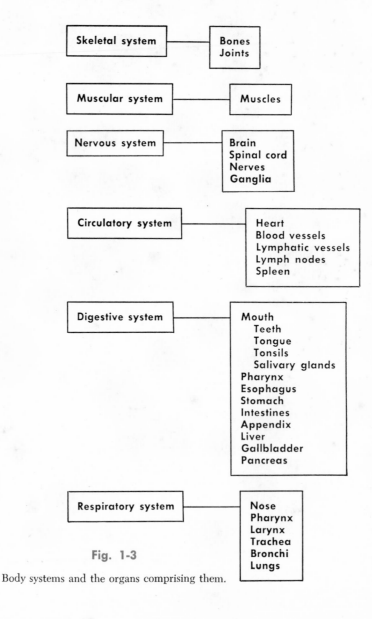

Fig. 1-3

Body systems and the organs comprising them.

About three hundred years ago, Robert Hooke looked through his microscope (one of the very early ones) at some plant material. What he saw must have surprised him. Instead of a single, much enlarged piece of plant material, he saw a group of many small pieces. They looked like miniature prison cells to him so that is what he called them—cells. This discovery of Hooke's that plants were made up of cells proved to be a foundation stone for modern biology. Thousands of individuals have examined thousands of plant and animal specimens since Hooke's time and have found all of them made up of cells. Today, therefore, biologists think of a *cell* as the unit of structure of living things, just as you might think of a brick as the unit of structure of a brick wall or of a brick house. In other words, you might think of a cell as a tiny "building block" for building living things. Some living things are so simple that all there is to them is just one cell; germs, for example (or to use

the scientific term, microorganisms), consist of only one cell. Some living things are so complex that they consist of billions of cells—the human body, for example.

When a living thing is made up of a great many cells, its cells are not all alike. Different kinds of cells perform different activities. Cells with one type of structure, for example, conduct impulses; cells with a different kind of structure contract; cells with still another type of structure secrete, and so on. Four main kinds of cells compose our bodies: epithelial cells, connective cells, muscle cells, and nerve cells. Many cells of one kind grouped together form a *tissue*. Thus, many epithelial cells grouped together form epithelial tissue, many muscle cells form muscle tissue, many connective tissue cells form connective tissue, and many nerve cells form nervous tissue. (The discussion of tissues begins on page 13.)

An *organ* is a structure composed of several kinds of tissues—often, of all four main kinds. Each of the tissues in an organ, in

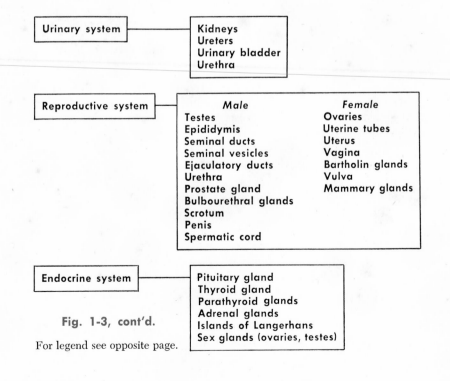

Fig. 1-3, cont'd.

For legend see opposite page.

turn, is composed of thousands of cells, An organ may also be defined as a structure that performs a more complex function for the body than any single cell or single tissue can achieve alone. You undoubtedly already know the names of most of the organs—stomach, heart, lungs, eyes, hands, feet, and many others. (See Figure 1-3.)

A *system* is the largest structural unit in the body. It consists of a group of organs that work together to perform a more complex function than any one organ can perform alone. Most anatomists seem to group organs into the following nine systems: skeletal, muscular, circulatory, digestive, respiratory, urinary, reproductive, endocrine, and nervous (Figure 1-3).

What structures make up the body? Briefly, cells, tissues, organs, and systems. Millions of cells form each tissue, two or more kinds of tissues form each organ, several organs form each system, and nine systems form the body.

The cell

Since every structure in the body is composed of cells, you need to know something about the structure and function of cells if you want to understand the structure and function of the body. All human cells are extremely tiny. Most of them measure only 1/2000th to 1/1000th of an inch across. Think of that—living units so small that 2,000 of them lined up would form a row only 1 inch long! Obviously you cannot see anything that tiny with your naked eye. Cells are not macroscopic, in other words. They are microscopic; only when they are magnified by a microscope can human cells be seen.

Cells differ in shape as well as in size. Some cells resemble tiny bricks; some are flat, more like the scales of a fish; some are long and slender like bits of thread; some are irregular in shape.

Cells have three main parts: cell membrane, cytoplasm, and nucleus (Figure 1-4). The *nucleus* is a round or oval struc-

ture in the central part of the cell. It is surrounded by *cytoplasm,* and the cytoplasm is enclosed by the *cell membrane.*

In addition to membrane, cytoplasm, and nucleus, cells, we now know, have numerous other smaller parts. Many of them are invisible even with the highest magnification a light microscope can provide (about a thousand times actual size). Light microscopes, therefore, did not reveal their existence, but electron microscopes did. With their magnifying power of many thousands of times, electron microscopes finally brought into man's view great numbers of ultrasmall cell structures never seen before. And now, not many years after their discovery, some of their names—mitochondria, ribosomes and lysosomes, for instance—have become familiar to almost everyone. Popular magazines* and even newspapers have carried articles about them.

Everything known so far about these structures sounds fantastic—more like a science fiction writer's dream than a scientist's facts. For one thing, despite their incredibly small size, they each have definite structure—notice, for instance, the shape of the mitochondria shown in Figure 1-4. These small cell structures also perform definite functions. In other words they function for the cell as organs function for the body, and for this reason they are called *organelles.* Figure 1-4 shows four different kinds of organelles in the cytoplasm of the cell: endoplasmic reticulum, ribosomes, mitochondria, and lysosomes. The following paragraphs describe their structure briefly. Their functions are discussed on page 11 and summarized in Table 1-1, page 13.

The *endoplasmic reticulum* consists of a network of extremely small, extremely thin-walled canals. They follow a rather twisted path through the cell's cytoplasm and eventually open on its surface. *Ribosomes,* as

*For example see: What causes inflammation and why it occurs, Time, June 16, 1967, p. 60.

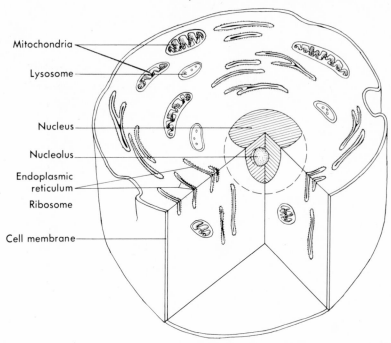

Mitochondria

Lysosome

Nucleus

Nucleolus

Endoplasmic
reticulum

Ribosome

Cell membrane

Fig. 1-4

Diagram of typical cell structure. Pie-shaped cut into the cell shows that cells are tridimensional, having thickness as well as length and width.

seen under the powerful magnification of an electron microscope, look like small granules. Great numbers of them lie along the endoplasmic reticulum, giving it the dotted appearance shown in Figure 1-4. *Mitochondria* are small sausage-shaped sacs. Actually, each mitochondrion consists of two sacs, one inside the other and each one composed of a very thin membrane. *Lysosomes* are also small membranous sacs. They contain powerful enzymes. (An *enzyme* is a protein compound that has the special ability to make one particular chemical reaction take place more rapidly. Every human cell contains hundreds of different enzymes.)

Cell structures and cell functions

If you think of a cell as a separate little individual, it may be easier for you to understand its functions. Amazing as it seems, each of the body's millions of cells does many of the same things that the whole body does. For instance, each cell breathes—oxygen moves into it and carbon dioxide

moves out of it. Each cell eats—food molecules move through its membrane into the cytoplasm. Each cell uses (metabolizes) food in two ways: it releases energy from some foods and from others it makes a number of complex compounds—enzymes and hormones, for instance. The process by which cells release energy from foods is called *catabolism*. The process by which cells convert foods into more complex compounds is called *anabolism*. The two processes, catabolism and anabolism, together constitute the process of metabolism. In the following paragraphs, we shall present first some basic information about catabolism and then some current ideas about which cell structures perform which cell functions.

Catabolism is a highly complex process. It consists of a large number of chemical reactions which go on continually inside every living cell. Not only that, but these

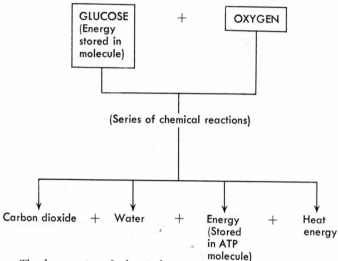

Fig. 1-5

Glucose catabolism. The long series of chemical reactions in glucose catabolism uses oxygen to convert glucose to carbon dioxide and water. But much more importantly these reactions transfer energy from the glucose molecule to ATP molecules where the energy is stored in a form that cells can use to do the work that keeps them and the body alive.

reactions also take place in a definite order or sequence. Each chemical reaction is catalyzed (accelerated) by its own special enzyme. Every cell, by the process of anabolism, makes hundreds of different enzymes. In fact, enzymes are the chief products of anabolism.

The main food catabolized by human cells is glucose. The chemical reactions of catabolism oxidize glucose. They change it into first one compound and then another, and finally, by adding oxygen they convert it to carbon dioxide and water. While these chemical changes are taking place, the energy stored in the glucose molecule is being released. Almost instantaneously, however, part of it is put back into storage—but in the molecules of another compound, namely, adenosine triphosphate (ATP). The rest of the energy originally stored in the glucose molecule is released as heat by the process of catabolism. ATP is one of the most important of all the compounds found in living things. Two characteristics of ATP energy account for this: it can be released almost instantaneously and when released it can be used directly to do cellular work. Neither of these qualities, on the other hand, applies to glucose energy. Its release occurs much more slowly, achieved by the long series of chemical reactions which make up the process of catabolism, and for some reason, energy released from glucose cannot be used directly to do cellular work. It must first be transferred to ATP molecules and be released from them. See Figure 1-5 for a summary of these changes.

Each cell structure performs definite functions for the cell. The *cell membrane,* for example, determines what substances move into or out of the cell. Some substances—notably oxygen, carbon dioxide, and water—move through the cell membrane under their own power by processes known as diffusion and osmosis. (See Glossary for definitions of these terms.) Other substances are carried through the membrane by means of energy supplied by the cell through catabolism. Such substances (mainly salts and foods) are said to be moved through the cell membrane by the process of active transport.

The *endoplasmic reticulum* is thought to serve as a miniature circulatory system for the cell. In other words, this network of canals appears to transport substances from one place to another within the cell. In secreting cells, compounds travel by way of the endoplasmic reticulum to reach the cell's surface and move out into the surrounding fluid.

Ribosomes are microscopic protein factories. They carry on the complex process of anabolism, particularly the making of enzymes and other proteins. From the ribosomes, these compounds presumably enter the canals of the endoplasmic reticulum for transportation to other parts of the cell.

Mitochondria serve as cellular powerhouses. Most of the chemical reactions of catabolism take place in mitochondria. Located in their membranous walls are the enzyme molecules that catalyze these various reactions. Since these reactions provide about ninety percent of a cell's energy supply, the organelles in which they occur—the mitochondria—well deserve their nickname "cellular powerhouses."

Lysosomes have been given two nicknames—"digestive bags" and "suicide bags." The reason for the first of these nicknames is that lysosomes contain enzymes that digest certain foods that enter the cell. Lysosomal enzymes can also digest other substances. For example, they can digest and thereby destroy microorganisms or other injurious substances which manage to invade the cell. Thus lysosomes can protect cells against destruction. But paradoxically, they can produce the opposite effect—lysosomal enzymes can actually kill cells. This happens if these powerful enzymes escape out of the normally tightly sealed lysosomes into the cell's cytoplasm. Then the lysosomal enzymes digest and destroy the cells themselves—hence the nickname "suicide bags." Presumably, lysosomes become transformed from digestive bags into suicide

bags under abnormal conditions. Some investigators suggest that this may happen when cells receive too little oxygen, when certain poisons attack them, or when inflammation develops.

Summarizing the functions performed by a cell's *nucleus* requires just two words, heredity and control. But to tell how the nucleus achieves these functions requires a long involved and still incomplete story. The key parts of the story have to do with two acids and numerous enzymes. You have probably seen the acids' initial names, DNA and RNA, many times. (DNA stands for deoxyribonucleic acid and RNA stands for ribonucleic acid.)

DNA molecules have a unique structure. Because their structure contains the crucial first clue for even partial understanding of heredity, we shall describe it briefly. To visualize a DNA molecule, first try to picture a ladder which has thousands of rungs. Imagine each rung consisting of either one or another of two pairs of molecules. A molecule of a compound called adenine may be paired with a molecule of a substance named thymine to form a rung of the DNA ladder. Or two other compounds, namely cytosine and guanine, may be paired together. Adenine, thymine, cytosine, and guanine are bases although DNA, the substance that they are part of, is an acid. Adenine joined to thymine is referred to as a "base-pair," and so, too, is cystosine joined to guanine. Fig. 1-6 illustrates the structure of an extremely small segment of an imaginary DNA molecule. It identifies the base-pairs composing four of its rungs as cytosine-guanine, thymine-adenine, guanine-cytosine, and adenine-thymine. If you look carefully at Figure 1-6, you will discover that two substances—a sugar and a phosphate—alternate to make up the sides of a DNA molecule. Notice that the base-pairs attach to the sugar, not the phosphate. Notice, too, that the D in DNA stands for the name of the sugar present in DNA—deoxyribose. Cell specialists estimate that

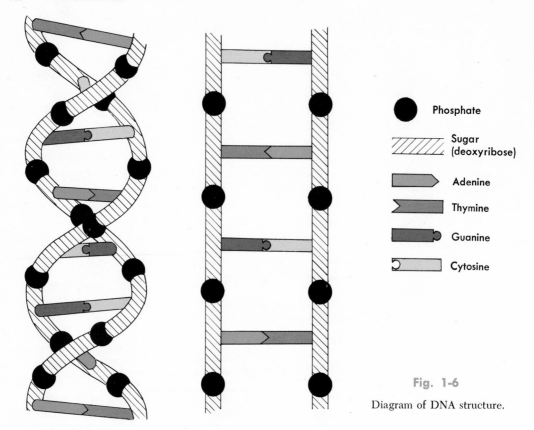

Phosphate

Sugar
(deoxyribose)

Adenine

Thymine

Guanine

Cytosine

Fig. 1-6

Diagram of DNA structure.

one human DNA molecule contains a sequence of some ten million base-pairs. You might, then, think of DNA molecules as microscopic ladders ten million steps long!

Another point should be mentioned about the structure of the DNA molecule. It is shaped like a twisted ladder rather than a straight one. (See Figure 1-6.) It spirals round and round thousands and thousands of times.

Chromosomes, which are very small structures in the cell nucleus, are composed mainly of DNA. Normal human cells—except mature sex cells—contain 23 pairs of chromosomes or a total of 46 chromosomes. Mature sex cells (ova and sperm) contain only half this number, that is, 23 single chromosomes.

Current ideas about how DNA functions to bring about heredity and control of cellular activities, told very briefly are as fol-

lows. The structure of a DNA molecule serves as a book of instructions. It gives directions for making enzymes. These instructions consist of the sequence in which one base-pair follows another in a DNA molecule. A long sequence of base-pairs (perhaps five hundred or a thousand of them, but exactly how many no one knows) is called a *gene.* Perhaps 20,000 genes, it is estimated, may make up one human DNA molecule! Genes in some way tell ribosomes what enzymes they are to make. How genes do this is far from fully known, but according to the popular "one gene-one enzyme" theory, each gene serves as a code for making one specific enzyme. Because enzymes catalyze the multitudes of chemical reactions continually taking place in living cells, they indirectly determine both cell functions and structure.

Each cell that is part of the body does

Table 1-1. Functions of some cell structures

Cell structures	Main functions
Cell membrane	Controls entrance and exit of substances into and out of cell
Endoplasmic reticulum	Transportation
Ribosomes	Anabolism
Mitochondria	Catabolism
Lysosomes	Digestion
Nucleus	Heredity and control

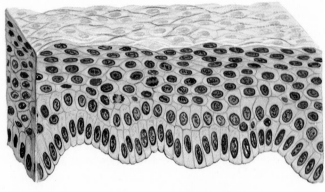

Fig. 1-7

Stratified squamous epithelium such as lines the mouth.

some special work for the rest of the body. Just as you, as an individual, do some kind of work for the community you live in, just as you specialize in nursing, teaching, selling, or some other kind of work, so each cell does some kind of work for the community it lives in—namely, the body. Some cells specialize in secreting, some in contracting, some in conducting impulses, and so on. Cells, in other words, serve the body as individual workers serve the community. The body, therefore, can only be as healthy as its cells. If enough of our cells become injured or diseased, we become sick, or even die.

Tissues

Epithelial tissue. Epithelial tissue is especially good at performing three functions for the body—protection, absorption, and secretion. The reason that it is good at these functions is because of its structure, that is, the shape and arrangement of its cells. Incidentally, here is a principle that you ought to notice and think about often when you are trying to learn about the body—namely, that structure determines function. This principle holds true too, of course, for nonliving things as well as living ones. (The structure of a sewing machine makes it possible for you to sew with it, but not to drive down a highway; to perform that function obviously you must have a machine with quite a different structure.)

There are several types of epithelial tis-

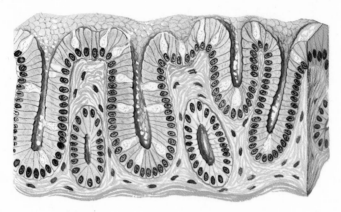

Fig. 1-8

Simple columnar epithelium with goblet cells, such as lines intestines.

sue. The structure of each one differs just enough from the structure of the others to make each type a specialist at a slightly different function. For example, in some parts of the body epithelial tissue consists of cells packed close together and arranged in several layers, one upon the other. (See Figure 1-7.) This arrangement of the cells makes this type of epithelial tissue—called *stratified squamous epithelium*—a specialist at protection. For instance, stratified squamous epithelial tissue protects the body against invasion by microorganisms. Most microorganisms cannot get through a barrier of stratified squamous tissue such as that which composes skin and mucous membrane surfaces. One way, therefore, that you can prevent infections is by taking good care of your skin. Don't let it become cracked from chapping. Guard against cuts and scratches—particularly good advice for nurses to remember.

In some parts of the body the main function of epithelial tissue is absorption. The cells of tissue in such areas are flat and arranged in a single layer. Substances therefore can pass through this tissue rapidly. This type of tissue is called *simple squamous epithelium*. (Squamous means shaped like fish scales.) It is the kind of tissue that forms the tiny air sacs in the lungs where oxygen is absorbed into the blood.

Secretion, the third function performed by epithelial tissue, is the chief function performed by a type of epithelial tissue called *simple columnar epithelium*. See Figure 1-8.

Connective tissue. Connective tissue gets its name from its main function, that of connecting one kind of tissue to another. Most types of connective tissue also furnish support for some structure. There are many different types of connective tissue. Areolar, fibrous, adipose, cartilage, bone, and hemopoietic tissue are some of the main ones. Of these, areolar and fibrous are the most abundant. Areolar tissue serves as the "glue" of the body, so it is found all over the body wherever one kind of tissue attaches to another. Fibrous tissue also connects various structures. For instance, it holds bones together at the joints and anchors muscles firmly to bones. Hemopoietic tissue is a highly important type of connective tissue. It forms the various kinds of blood cells, and these, in turn, perform several vital functions.

The structure of connective tissue differs from that of other tissues primarily in the amount of material found between its cells (in the amount of intercellular substance, that is) and in the many different varieties of its cells. Fibroblasts, macrophages, plasma cells, osteocytes, and fat cells are

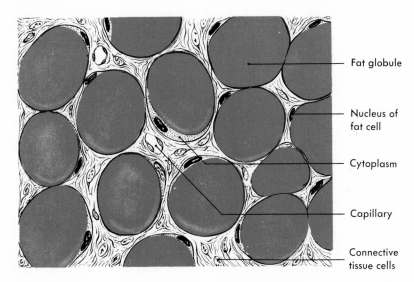

Fat globule

Nucleus of
fat cell

Cytoplasm

Capillary

Connective
tissue cells

Fig. 1-9

the names of a few of these different types of connective tissue cells. Figure 1-9 illustrates fat cells.

Muscle tissue. The special function of muscle tissue is contraction. There are three different types of muscle tissue: the tissue that makes up the muscles that are attached to bones, the tissue that helps form the walls of many tube-shaped organs (such as blood vessels and intestines), and the tissue that composes the heart. For a more detailed discussion, see pages 33 and 34.

Nervous tissue. Nervous tissue cells are discussed on page 48.

Organs

Because the skin is one of the most wonderful organs of the body, we shall make it the subject of our discussion of organs.

The skin. By virtue of size alone the skin occupies a position of great importance. It is one of the largest organs in the body despite the fact that it is only paper-thin in some places and not much thicker in the thickest places. Architecturally the skin is a marvel. Consider the incredible number of structures fitted into an area no bigger than your little fingernail: several dozens of

Fat cells. In fat cells, which make up adipose tissue, a fat droplet occupies nearly the entire area of the cell. The cytoplasm and nucleus are forced to the periphery of the cell.

sweat glands, hundreds of nerve endings, yards of tiny blood vessels, numerous oil glands and hairs, and literally thousands of cells. The skin performs lifesaving functions. Not only does it protect our bodies against invasion by deadly numbers of microorganisms, but it also plays an important part in maintaining normal body temperature and water balance.

Two kinds of tissue compose the skin (Figure 1-10). Stratified epithelial tissue makes up the outer or surface layer known as the *epidermis;* connective tissue makes up the thicker underlying layer called the *dermis.* As you may recall, the cells of stratified epithelial tissue are closely packed together and are arranged in several layers. Only the cells of the innermost layer of the epidermis reproduce themselves. As they increase in number, they move up toward the surface and dislodge the dead horny cells of the outer layer. These flake off by the thousands onto our clothes, into our

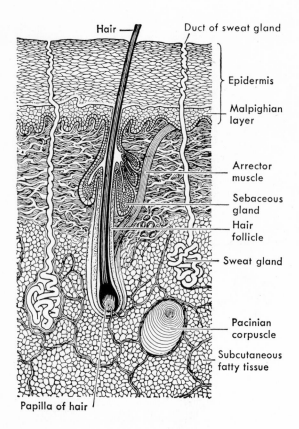

Hair —
Duct of sweat gland

Epidermis

Malpighian layer

Arrector muscle

Sebaceous gland

Hair follicle

Sweat gland

Pacinian corpuscle

Subcutaneous fatty tissue

Papilla of hair

Fig. 1-10

Diagram showing the structure of skin as seen in a longitudinal section magnified greatly by a microscope. The dermis, which is not labeled, is the thick layer shown between the epidermis and the subcutaneous fatty tissue. (Modified from Cunningham; from Tuttle, W. W., and Schottelius, Byron A.: Textbook of physiology, St. Louis, 1965, The C. V. Mosby Co.)

bath water, onto things we handle. Millions of epithelial cells are produced daily to replace the millions shed—just one example of work our bodies do without our knowing it, even when they seem to be resting.

Have you ever wondered what gives color to your skin, or why its color differs from time to time? (At different seasons of the year, for example, or when you are sick, or angry, or frightened, or too warm, or too cold?) Skin color changes because of two factors: (1) the amount of a pigment, called melanin, in the epidermis; and (2) the amount of blood in the dermis and also the kind of blood (that is, whether it contains much or little oxygen). Skin tans after long exposure to sunlight because ultraviolet rays increase the formation of melanin. It reddens when heat or emotions cause skin capillaries to fill with more

blood. It pales when sickness or cold decreases the amount of blood. It becomes bluish when the blood contains too little oxygen. A nurse or doctor can use this knowledge to judge quickly whether a normal amount of blood is circulating through the skin and whether blood is carrying enough oxygen to body cells.

The deep layer of the skin, the dermis, has several structural points of interest. Look at the palms of your hands and you will see one of these—the ridges and grooves that make possible fingerprinting as a means of identification. The epidermis follows the contours of ridges present in the dermis. These develop sometime before birth. Not only is their pattern unique for each individual, but also it never changes except to grow larger—two facts that explain why our fingerprints or footprints positively identify us. Many hospitals today identify each newborn baby by foot-printing it almost as soon as it is born.

Connective tissue composes the dermis. Instead of cells being crowded close together like the epithelial cells of the epidermis, they are scattered far apart, with many fibers in between. Some of the fibers are tough and strong (collagenous or white fibers), and others are stretchable and elastic (elastic fibers). Some of the cells of the dermis store fat. The wrinkles that come with age are thought to be due partly to a decrease in skin fat and partly to a decrease in the number of elastic fibers. A rich network of capillaries lies in the dermis just under the epidermis. The epidermis itself, contains no blood vessels. Surgeons sometimes use the knowledge of this anatomical fact to treat certain skin conditions by cutting such thin shavings from the surface of the skin that they remove only the epidermis. As you would expect, this is a bloodless operation. In addition to blood vessels, the dermis contains the following structures: nerves, roots of the hairs, oil glands, and sweat glands.

Hair. If you have ever looked closely at a newborn baby, you probably noticed its soft downy hair. If the baby was born prematurely, you may have found the same type hair (called *"lanugo"*, from the Latin word that means down or wool) all over its tiny body. Coarser hair soon replaces the lanugo of the scalp and eyebrows, but new hair on the rest of the body generally stays delicate and downlike. Hair grows from a cluster of epithelial cells at the bottom of the hair follicle. As long as these cells remain alive, new hair will replace any that is cut or plucked. Contrary to popular belief, frequent cutting or shaving does not make hair grow faster or become coarser, because neither process affects the epithelial cells that form the hairs, embedded as they are down in the scalp.

Oil glands. Wherever hairs grow, oil or sebaceous glands also grow. Their tiny ducts open into hair follicles so that their secretion (sebum) lubricates the hair as well as the skin. Someone aptly described sebum as "nature's cold cream." Besides keeping the hair and skin soft and pliant, this natural oil also helps preserve the normal water content of the body by hindering water loss from the skin by evaporation. This, in turn, helps preserve normal body temperature. Because evaporation has a cooling effect and because the presence of oil on the skin slows down evaporation, it also slows down cooling.

A small involuntary muscle attaches to the side of each hair follicle just below its oil gland and extends upward on a slant to attach to the skin. When it contracts, as it often does when we are cold or emotionally upset, it produces several effects. It squeezes on the oil gland with the same result that you would produce if you squeezed on an oil can—a tiny bit of oil squirts out of each follicle onto the skin. Also, as the muscle contracts it simultaneously pulls on its two points of attachment—that is, up on a hair follicle but down on a point of skin. This produces little raised places (goose pimples) between the depressed points of the skin and, at the same time, pulls the hairs up more or less straight. Incidentally, the latter fact accounts for the name of these muscles, arrectores pilorum (Latin for "erectors of the hair"). We unconsciously recognize these facts in pictorial expressions such as "I was so frightened my hair stood on end" and "She was so angry she was fairly bristling."

Sweat glands. The pinpoint-sized openings on the skin that you probably think of as pores are really outlets of small ducts from the sweat glands. The glands themselves lie in the subcutaneous tissue just under the dermis, and their ducts spiral up through the epidermis to the surface. There are great numbers of them. For instance, the palms of your hands are estimated to have about 3,000 sweat glands per square inch! The soles of the feet, the forehead, and the axillae (armpits) also contain a great many of them, more than most parts. The size of sweat glands (they are microscopic structures) belies their importance. They play an important part in the two vital functions of maintaining water balance and body temperature; more will be said about these later. Sweat glands also play a part, but a minor one, in excreting certain wastes from the body.

Systems

The nine systems are discussed in the remaining chapters of this book.

Outline summary—General plan and structural units of the body

GENERAL PLAN

Thoracic cavity

Contains lungs, heart, trachea, esophagus, and thymus gland

Abdominopelvic cavity

1. Abdominal cavity contains stomach, intestines, liver, gallbladder, pancreas, and spleen
2. Pelvic cavity contains reproductive organs, urinary bladder, lowest part of intestine

Cranial cavity

Contains brain

Spinal cavity

Contains spinal cord

STRUCTURAL UNITS

1. Cells—structural units of all living things
2. Tissues—groups of like cells
3. Organs—composed of more than one kind of tissues; perform more complex functions than a single tissue
4. Systems—groups of organs; perform more complex functions than a single organ

The cell

1. Structure
    a. Size—microscopic but varies with type of cell
    b. Shape—varies with type of cell
    c. Main parts—nucleus, cytoplasm, and cell membrane
    d. Some other cell structures—endoplasmic reticulum, ribosomes, mitochondria, and lysosomes
2. Functions
    a. Each cell performs functions necessary for its own life, such as respiration, ingestion of food and metabolism (catabolism plus anabolism)
    b. Different cell structures perform different functions—see Table 1-1, page 13
    c. Each cell performs some special function for body as a whole, such as contraction by muscle cells, conduction of impulses by nerve cells

Tissues

1. Epithelial tissue
    a. Function—protection, absorption, and secretion
    b. Types—several; for example, stratified and simple squamous epithelial tissue
2. Connective tissue
    a. Functions—connection, support, and blood cell formation
    b. Types—numerous; for example, areolar, fibrous, adipose, cartilage, bone, blood, and hemopoietic tissues
3. Muscle tissue
    a. Function—contraction
    b. Types—discussed in Chapter 3
4. Nervous tissue—discussed in Chapter 4

Organs

Skin

Composed of two main layers

1. Epidermis—outer layer of stratified squamous epithelial tissue; dead surface cells continually flake off; contains pigment (melanin)
2. Dermis—underlying layer of connective tissue; ridges and grooves in dermis form pattern unique to each individual—basis of fingerprinting; blood vessels nerves, hair roots, oil glands, sweat glands in dermis; fat stored in dermis
3. Hair—grows from cluster of epithelial cells at bottom of hair follicle deep in dermis
4. Oil glands—secrete oil into hair follicles; oil keeps hair and skin soft; also helps prevent loss of water and heat from skin
5. Sweat glands—openings from their ducts are "pores" of skin; amount of sweat secreted helps control amount of water and heat lost from body

Systems

See chapters 2 to 10

Review questions—General plan and structural units of the body

1. In what cavity is each of the following found?

appendix	liver
brain	lungs
esophagus	pancreas
gallbladder	spinal cord
heart	spleen
intestines	urinary bladder

2. Explain what the following terms mean: cells, organs, systems, tissues.
3. What is the primary function of the cell membrane?
4. What and where are the following? What functions do they perform?   *R. 11*

endoplasmic	mitochondria
reticulum	ribosomes
lysosomes	

DETERMINES WHAT SUBSTANCES MOVE INTO OR OUT OF THE CELL.

5. What is the full name of the now famous acid found in the nuclei of cells?
6. This acid makes up the major part of what microscopic structures?
7. Very briefly, what function does DNA perform?
8. Give a modern definition for the word gene.
9. What functions does epithelial tissue specialize in? Muscle tissue? Connective tissue?
10. Describe briefly the epidermis and dermis.
11. Why is oil and sweat secretion by the skin glands important to the body?

How the body stands erect and moves

II

The skeletal system

Before studying this chapter, let your imagination go for a few minutes. Think of your bones suddenly turning soft, into a material, say, of the consistency of a piece of liver. Suppose you were standing when this change took place. What do you see happening? Suppose you fell and struck your head. Would you bruise your scalp only or your brain as well? Now change your mental picture. Think this time of a single piece of wood cut in the shape of the human body. Try to visualize this structure sitting down, or bending over, or throwing a ball, or walking across a room. Can you see it moving any of its parts or making any one of the hundreds of movements that you make every day without even giving them a thought?

FUNCTIONS

The images you have just called up point out three important functions of the skeletal system: it supports and gives shape to the body, it protects internal organs, and it makes movements possible. The skeletal system also contributes another function—one which you probably would not guess—the formation of blood cells. The process of blood cell formation, or hemopoiesis, is discussed on page 68.

How the skeletal system makes movements possible will be considered partly in this chapter in the discussion of joints and partly in the next chapter in the discussion of muscle actions. We shall consider now the general plan of the skeleton, the individual bones, the structure of a typical long

23

bone, and the structure and function of joints.

GENERAL PLAN

The human skeleton has two divisions: the *axial skeleton* and *appendicular skeleton*. The main parts of these two divisions are listed in Table 2-1.

The skull

Twenty-eight bones compose the skull. You will probably want to learn their names and find out what part of the skull each one forms. Their names are given in Table 2-2. Find as many of them as you can on Figure 2-1. Feel their outlines on your own body where possible. Examine them on a

Table 2-1. Main parts of the skeleton

Axial skeleton	*Appendicular skeleton*
Skull	Upper extremities
Cranium	Shoulder girdle
Face	Arms
Spine	Hands
Vertebrae	Lower extremities
Thorax	Hip girdle
Ribs	Legs
Sternum	Feet
Ear bones	
Hyoid bone	

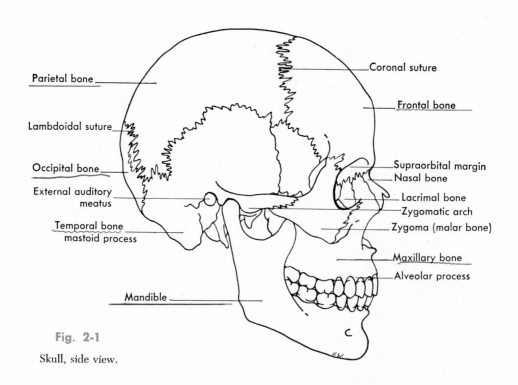

Fig. 2-1

Skull, side view.

Table 2-2. Bones of the skeleton

Name	Number	Description
Cranial bones		
Frontal	1	Forehead bone; also forms front part of floor of cranium and most of upper part of eye sockets; cavity inside bone above upper margins of eye sockets (orbits) called *frontal sinus;* lined with mucous membrane
Parietal	2	Form bulging topsides of cranium
Temporal	2	Form lower sides of cranium; contain *middle and inner ear structures; mastoid sinuses* are mucous-lined spaces in *mastoid process,* the protuberance behind ear; *external auditory canal* is tube into temporal bone
Occipital	1	Forms back of skull; spinal cord enters cranium through large hole (*foramen magnum*) in occipital bone
Sphenoid	1	Forms central part of floor of cranium; pituitary gland located in small depression in sphenoid called *Turk's saddle* or sella turcica
Ethmoid	1	Complicated bone that helps form floor of cranium, side walls and roof of nose and part of its middle partition (nasal septum), and part of orbit; contains honeycomb-like spaces, the *ethmoid sinuses; superior and middle turbinate bones* (conchae) are projections of ethmoid bone; they form "ledges" along side wall of each nasal cavity
Face bones		
Nasal	2	Small bones that form upper part of bridge of nose
Maxillary	2	Upper jawbones; also help form roof of mouth, floor, and side walls of nose and floor of orbit; large cavity in maxillary bone is *maxillary sinus*
Zygoma (malar)	2	Cheek bones; also help form orbit
Mandible	1	Lower jawbone
Lacrimal	2	Small bone; helps form medial wall of eye socket and side wall of nasal cavity
Palatine	2	Form back part of roof of mouth and floor and side walls of nose and part of floor of orbit
Inferior turbinate	2	Form curved "ledge" along inside of side wall of nose, below middle turbinate
Vomer	1	Forms lower, back part of nasal septum
Ear bones		
Malleus	2	Malleus, incus, and stapes are tiny bones in middle ear cavity in temporal bone; name, malleus, means hammer—shape of bone
Incus	2	Incus means anvil—shape of bone
Stapes	2	Stapes means stirrup—shape of bone
Hyoid bone	1	U-shaped bone in neck at base of tongue
Vertebral column		
Cervical vertebrae	7	Upper seven vertebrae, in neck region; first cervical vertebra called *atlas;* second called *axis*
Thoracic vertebrae	12	Next twelve vertebrae; ribs attach to these
Lumbar vertebrae	5	Next five vertebrae; those in small of back
Sacrum	1	In child, five separate vertebrae; in adult, fused into one
Coccyx	1	In child, three to five separate vertebrae; in adult, fused into one

Continued.

Table 2-2. Bones of the skeleton—cont'd

Name	Number	Description
Thorax		
True ribs	14	Upper seven pairs; attached to sternum by way of *costal cartilages*
False ribs	10	Lower five pairs; lowest two pairs do not attach to sternum, therefore, called *floating ribs;* next three pairs attach to sternum by way of costal cartilage of seventh ribs
Sternum	1	Breast bone; shaped like a dagger; piece of cartilage at lower end of bone called *xiphoid process*
Upper extremities		
Clavicle	2	Collar bones; only joints between shoulder girdle and axial skeleton are those between each clavicle and sternum
Scapula	2	Shoulder bones; scapula plus clavicle forms *shoulder girdle; acromion process*—tip of shoulder that forms joint with clavicle; *glenoid cavity*—arm socket
Humerus	2	Upper arm bone
Radius	2	Bone on thumb side of lower arm
Ulna	2	Bone on little finger side of lower arm; *olecranon process*—projection of ulna known as the elbow or "funny bone"
Carpal bones	16	Irregular bones at upper end of hand; anatomical wrist
Metacarpals	10	Form framework of palm of hand
Phalanges	28	Finger bones; three in each finger, two in each thumb
Lower extremities		
Pelvic bones	2	Hip bones; *ilium*—upper flaring part of pelvic bone; *ischium*—lower back part; *pubic bone*—lower, front part; acetabulum—hip socket; *symphysis pubis*—joint in midline between two pubic bones; pelvic inlet—opening into *true pelvis*, or pelvic cavity; if pelvic inlet is misshapen or too small, infant skull cannot enter true pelvis for natural birth
Femur	2	Thigh or upper leg bones; *head of femur*—ball-shaped upper end of bone; fits into acetabulum
Patella	2	Kneecap
Tibia	2	Shin bone; *medial malleolus*—rounded projection at lower end of tibia commonly called inner ankle bone
Fibula	2	Long slender bone of lateral side of lower leg; *lateral malleolus*—rounded projection at lower end of fibula commonly called outer ankle bone
Tarsal bones	14	Form heel and back part of foot; anatomic ankle
Metatarsals	10	Form part of foot to which toes attach; tarsal and metatarsal bones so arranged that they form three arches in foot: *the inner longitudinal arch* and the *outer longitudinal arch*, both of which extend from front to back of foot, and transverse or *metatarsal arch* that extends across foot
Phalanges	28	Toe bones; three in each toe, except great toes, where there are two
Total	206	

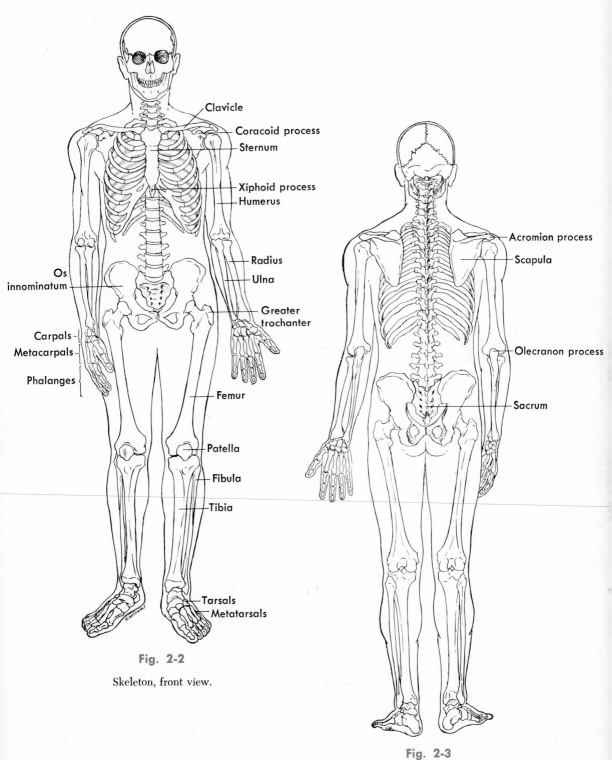

Clavicle
Coracoid process
Sternum
Xiphoid process
Humerus
Radius
Ulna
Greater trochanter
Os innominatum
Carpals
Metacarpals
Phalanges
Femur
Patella
Fibula
Tibia
Tarsals
Metatarsals

Acromion process
Scapula
Olecranon process
Sacrum

Fig. 2-2

Skeleton, front view.

Fig. 2-3

Skeleton, back view.

skeleton if you have access to one. Look in Table 2-2 for the answers to question 9 on page 32.

"My sinuses give me so much trouble." Have you ever heard this complaint or perhaps uttered it yourself? *Sinuses* are spaces or cavities inside some of the cranial bones. Four pairs of them (those in the frontal, maxillary, sphenoid, and ethmoid bones) have openings into the nose and so are referred to as *paranasal sinuses*. Sinuses give trouble when the mucous membrane that lines them becomes inflamed, swollen, and painful. For example, inflammation in the frontal sinus *(frontal sinusitis)* often starts from a common cold. (The letters "itis" added to a word mean inflammation of.)

The spine (vertebral column)

The term "vertebral column" may suggest a mental picture of the spine as a single long bone shaped like a column in a building, but this is far from true. The vertebral column consists of a series of separate bones (vertebrae) connected in such a way that they form a flexible curved rod. Different sections of the spine have different names: cervical region, thoracic region, lumbar region, sacrum, and coccyx. These are defined in Table 2-2.

The curved shape of the spine fulfills an important function—it makes the spine exceptionally strong. This is because a curved structure is stronger than a straight one. (The next time you pass a bridge look to see whether its supports form a curve.) Obviously, the spine needs to be an exceptionally strong structure. It supports the head balanced on top of it, the ribs and internal organs suspended from it in front, and the hips and legs attached to it below. A new born baby's spine forms a continuous curve from top to bottom. This changes, however, as the baby learns to hold his head up. Then a reverse curve develops in the neck (cervical region). Later, as the baby learns to stand, another reverse curve

appears in the small of his back (lumbar region).

Disease or poor posture often causes the lumbar curve to become abnormally exaggerated, a condition commonly called "swayback" or, technically, *lordosis*. Another abnormal curvature is *kyphosis*, known to most of us as "hunchback."

The thorax

Twelve pairs of ribs, the sternum (breast bone), and the thoracic vertebrae form the bony cage known as the thorax or chest. The terms "true ribs," "false ribs," "floating ribs," and "costal cartilage" are explained in Table 2-2.

The upper extremities

Stop and think a minute of how many ways you can move your shoulders, arms, and hands. The framework of these parts consist of a good many bones—thirty-two in each upper extremity, to be exact. If instead, only one bone ran the length of your arm and hand, how many movements do you think you could make with them? Could you, for example, bend your elbow? Could you move your hands? Table 2-2 names and describes the bones of the upper extremities.

The lower extremities

Like the upper extremities, the lower extremities can perform various movements. For this reason they, also, have a framework of many bones connected by movable joints. The names of the bones and their descriptions are given in Table 2-2.

STRUCTURE OF LONG BONES

Figures 2-4 and 2-5 will help you learn the names of the main parts of a long bone. They are as follows:

1. *Diaphysis,* or shaft—made of hard compact bone; it is a hollowed-out cylinder filled with yellow bone marrow;
2. *Epiphyses,* or the ends of the bone—

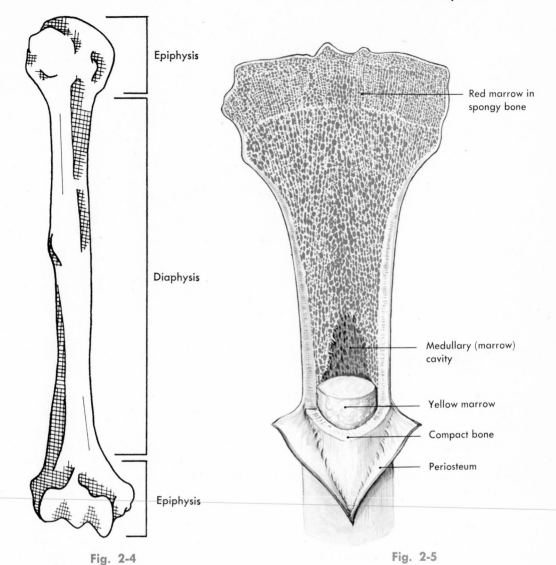

Epiphysis

Diaphysis

Epiphysis

Red marrow in
spongy bone

Medullary (marrow)
cavity

Yellow marrow

Compact bone

Periosteum

Fig. 2-4

Divisions of a long bone.

Fig. 2-5

Cutaway section of a long bone.

red bone marrow fills in the small
spaces in the spongy bone composing the epiphyses;

3. *Articular cartilage*—a thin layer of cartilage covering each epiphysis; serves
much the same purpose that a small
rubber cushion would if it were
placed over the ends of bones where
they come together to form a joint;

4. *Periosteum*—a strong fibrous mem-

brane that covers a long bone except
at its joint surfaces where it is covered
by articular cartilage;

5. *Medullary cavity*—the hollow area inside the shaft of a bone; it contains
yellow bone marrow.

JOINTS (ARTICULATIONS)

Every bone in the body, except one, connects to at least one other bone. In other

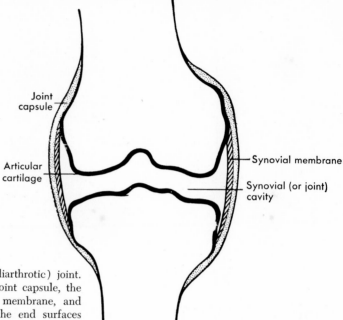

Joint capsule

Articular cartilage

Synovial membrane

Synovial (or joint) cavity

Fig. 2-6

Structure of a freely moveable (diarthrotic) joint. Note these typical features: the joint capsule, the joint cavity lined with synovial membrane, and the articular cartilage covering the end surfaces of the bones within the joint capsule.

words, every bone but one forms a joint with some other bone. (The exception is the hyoid bone in the neck, to which the tongue anchors.) Most of us never think much about our joints unless something goes wrong with them so that they do not function properly. Then their tremendous importance becomes painfully clear. Joints perform two functions: they hold our bones together securely, and, at the same time, they make it possible for movement to occur between the bones—between most of them, that is. In a few places, though, joints are constructed so that no movement can take place. Most joints in the skull, for example, are immovable. A few joints allow only very slight movement, but by far the majority of them allow considerable movement—sometimes in many directions and sometimes in only one or two directions. Try, for example, to move your arm at your shoulder joint in as many directions as you can. Try to do the same thing at your elbow joint. Now examine the shape of the bones at each of these joints on a skeleton or in Figures 2-2 and 2-3. Do you see why

you cannot move your arm at your elbow in nearly as many directions as you can at your shoulder?

Freely movable joints are all made alike in certain ways. For example, all of them have a joint capsule, a joint cavity lined with smooth, slippery synovial membrane, and a layer of cartilage over the ends of the two joining bones (Figure 2-6). The *joint capsule* is made of the body's strongest toughest material—fibrous connective tissue. It fits over the ends of the two bones something like a sleeve. Because it attaches firmly to the shaft of each bone to form its covering (called the periosteum; "peri" means around and "osteum" means bone), the joint capsule holds the bones together securely but at the same time permits movement at the joint. The structure of the joint capsule, in other words, helps make possible the joint's function.

Ligaments (cords or bands of the same strong fibrous tissue) also grow out of the periosteum and lash the two bones together even more firmly.

The layer of *articular cartilage* over the

joint ends of bones acts like a rubber heel on a shoe—it absorbs jolts. *Synovial membrane* secretes a lubricating fluid (synovial fluid) that allows easier movement with less friction.

Because of the structure of freely movable joints, we can move them in a variety of ways. These will be discussed with muscle action in the next chapter.

Various things can go wrong with our joints. Arthritis or joint inflammation, for example, cripples hundreds each year. Medical science has not yet learned how to defeat this powerful enemy even though it has developed some valuable weapons to use against it. The hormone hydrocortisone is one of these. Dramatic relief often follows the injection of this preparation into a joint cavity.

Advancing years take their toll of the joints by bringing degenerative changes in them. Especially is this true for the weight-bearing joints and for the joints of overweight individuals. Degenerative joint changes lead to the complaints we hear so often from older people: "My joints are getting so stiff, I even have trouble getting up out of a chair" or "I'm not nearly as spry as I used to be; I can't move around so easily." Two of the commonest kinds of joint changes are adhesions that form between the cartilage-covered joint surfaces of the bones and extra bone that grows along the joint edges (called "spurs" and "marginal lipping").

MISCELLANEOUS FACTS

A man's skeleton and a woman's differ somewhat. For instance, a man's skeleton is usually larger than a woman's, and his hip bones are generally shaped differently. In most men, the hip or pelvic bones form a narrow deep funnel-shaped pelvis. In most women, on the other hand, they form a broad, shallow, basin-shaped pelvis, one wide enough for a baby's body to pass through in its birth journey.

Have you ever noticed that a newborn baby's head seems larger for the size of its body than an adult's head does? This is one of several differences between a baby's and an adult's skeleton. A newborn baby's head is about ¼ as long as its whole body, whereas an adult's head is only about ⅛ of its total height. A baby's chest is round, and an adult's is oval. A baby's face appears small and its cranium large; an adult's face and cranium look nearly the same size.

At birth the skeleton is unfinished in that some of the bones still consist partly of cartilage or fibrous tissue. Familiar examples of this are the soft spots (fontanels) in a baby's head. Another example is the cartilage between the shaft of a long bone and its ends. This is known as the epiphyseal cartilage, and it remains as long as bones are still growing. Epiphyseal cartilage can be seen in x-ray pictures—a convenient fact if a doctor wants to know whether a child is going to grow anymore. No epiphyseal cartilage means no more growth in bone length.

Bones grow normally only if the blood supplies them with plenty of minerals (calcium and phosphorus especially) and vitamin D. Therefore, if a child's diet contains too little calcium or vitamin D, he soon suffers from the disease known as rickets. His bones do not harden, that is, calcify normally. As a result his leg bones bend from bearing the weight of his body and he becomes bowlegged.

Bone disorders occur even more often in elderly people than in children. For instance old people's bones usually become more porous. This change may make the bones so weak that they break under the body's weight—a fact that probably explains many of the broken hips in old people. Exactly what causes osteoporosis (condition in which bones are more porous than normal) no one yet knows for sure, but too little exercise and too small amounts of sex hormones presumably have a good deal to do with it. Consider these facts as evidence: from the menopause on, women characteristically have an estrogen deficiency and exercise sparingly. They also—

most of them, that is—have some degree of osteoporosis. In short, menopause, deficient estrogens, deficient exercise, and osteoporosis more often than not go together. Osteoporosis is common also in older men.

Outline summary—The skeletal system

FUNCTIONS

1. Supports and gives shape to body
2. Protects internal organs
3. Helps make movements possible
4. Forms blood cells

GENERAL PLAN

Consists of two main divisions, several subdivisions

Axial skeleton

1. Skull
2. Spine
3. Thorax
4. Ear bones
5. Hyoid bone

Appendicular skeleton

1. Upper extremities, including shoulder girdle
2. Lower extremities, including hip girdle

LOCATION AND DESCRIPTION OF BONES

See Figures 2-1, 2-2, 2-3, and Table 2-2

STRUCTURE OF LONG BONES

See Figures 2-4, 2-5

JOINTS (ARTICULATIONS)

1. Functions—hold bones together securely; make possible movements
2. Structures of freely movable joints
   a. Joint capsule and ligaments hold joining bones together but permit movement at joint
   b. Articular cartilage—covers joint ends of bones and absorbs jolts
   c. Synovial membrane—lines joint capsule and secretes lubricating fluid
   d. Joint cavity—space between joint ends of bones

3. Joint disorders—arthritis and age degeneration are two of commonest

MISCELLANEOUS FACTS

1. Male and female skeletons differ; for example, male hip bones form narrow deep funnel, whereas female hip bones form broad shallow basin
2. Infant and adult skeletons differ; for example, infant head forms larger proportion of body height than adult head does
3. Bone growth depends on minerals, vitamin D, and blood supply to bone
4. Bone disorders common in elderly people

Review questions—The skeletal system

1. What functions does the skeleton system perform for the body?
2. Name the bones of the skull.
3. Name the small bones in the middle ear.
4. What bones make up the shoulder girdle?
5. What bones make up the arms and hands?
6. What bones make up the lower extermities?
7. What functions do joints perform?
8. Describe the structure of a freely movable joint.
9. Locate and briefly describe each of the following bones:

clavicle	patella
ethmoid	radius
femur	scapula
frontal	sphenoid
humerus	sternum
ilium	tibia
mandible	turbinates
metacarpals	zygomatic
parietal	ulna

10. Describe one marked difference between a male and a female adult skeleton.
11. Describe some differences between an infant and an adult skeleton.
12. Is it possible to tell whether a child is going to grow any taller? If so, how can a doctor tell this?
13. Why are calcium and vitamin D so important in a child's diet?
14. What structural changes frequently occur in older people's joints?

III

The muscular system

When we speak of the muscular system, we mean the more than 500 muscles by which we move in many ways, varying in complexity from blinking an eye or smiling to climbing a mountain or ski-jumping. Not many of our body structures can claim as great importance for happy useful living as can our voluntary muscles, and only a few can boast of greater importance for life itself. A great deal is known about muscles—enough, in fact, to fill several books the size of this one. So in this chapter we shall try to present only information that practical nurses and other beginning students of the body will find useful to know. The plan is this—to investigate first the different types of muscle tissue, then to note some general facts about the structure and function of skeletal muscles and to present some specific facts about certain key muscles and about posture, and finally to consider some muscle disorders.

MUSCLE TISSUE

If you weigh 120 pounds, about 50 pounds or more of your weight comes from your muscles, from the "red meat" attached to your bones. Under the microscope these muscles appear as bundles of fine threads with many crosswise stripes. Each fine thread is a muscle cell or, as it is usually called, a *muscle fiber*. This type of muscle tissue goes by three names: *striated muscle*—because of its cross stripes or stria, *skeletal muscle*—because it attaches to bone, and *voluntary muscle*—because its contractions can be controlled voluntarily.

Beside skeletal muscle, the body also contains two other kinds of muscle tissue, and each of these also has more than one name. One kind is called *branching muscle*—because its cells seem to branch into each other—or *cardiac muscle*—because it composes the bulk of the heart. The third kind of muscle tissue consists of slender tapered

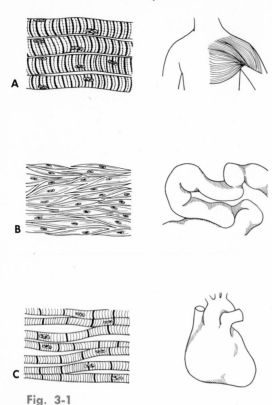

Fig. 3-1

Muscle tissue of the human body. **A,** Striated or skeletal muscle tissue. On the left is a microscopic view showing cross striations and multinuclei per cell; on the right, a macroscopic view of skeletal muscle organs. **B,** Nonstriated or smooth muscle tissue. On the left, is a microscopic view; on the right, a loop of intestine, one of many internal organs whose walls contain smooth muscle. **C,** Branching or cardiac muscle tissue. On the left is a microscopic view showing cross striations and branching cells; on the right, the heart, the only organ made of cardiac muscle tissue.

cells that have no cross stripes. *Nonstriated muscle* and *smooth muscle,* therefore, are the names given to this tissue. It is also called *involuntary muscle*—for the obvious reason that most people cannot voluntarily control the contraction of this type of muscle. Some rather startling exceptions to this rule do exist, however. Smooth muscle forms an important part of blood vessel walls and of many internal organs (viscera), so another

name for it is *visceral muscle.* (See Figure 3-1).

GENERAL FACTS ABOUT SKELETAL MUSCLES
Functions

Two words—movement and posture—should come to mind when we think of skeletal muscles, for these are the muscles whose contractions move us about in almost endless ways. But, paradoxically, they are also the ones that hold us motionless in various positions, that maintain our postures, in other words. (The word *posture* means position—position of the body as a whole or position of any of its parts.)

Most skeletal muscles attach to two bones joined together by a movable joint. In other words, most muscles extend from one bone across a joint to another bone. Also, one of the two bones moves less easily than the other. The muscle's attachment to this more stationary bone is called its *origin.* Its attachment to the more movable bone is its *insertion.* The rest of the muscle, that is, all of it except its two ends, is called the *body* of the muscle. Strong cord-like structures made of dense fibrous connective tissue, which are called *tendons,* attach muscles to bones. Any emergency room nurse or doctor sees many tendon injuries—severed tendons, for instance, and tendons pulled away from bones.

Small sacs called *bursae* and made of connective tissue lined with synovial membrane are located between tendons and bones and around joints. They serve as little cushions, decreasing the pressure on various structures during movement. *Bursitis* is an inflammation of a bursa.

When a muscle decreases in length as it contracts, the shortening of its fibers necessarily exerts a pull on the bones the muscle attaches to. As you would expect, this pulling force causes the more easily moved bone to move—causes the insertion bone to move, in other words. The origin bone stays put, holding firm while the insertion bone is pulled toward it. To summarize, it is always the insertion bone that moves when a muscle contracts, and it always moves nearer to the origin bone.

Another fact important to remember is that muscles generally work in teams, not singly. Several muscles contract at the same time to produce almost any movement you can think of. Of the muscles contracting simultaneously, the one mainly responsible for producing the movement is called the *prime mover* for that movement. The other muscles, those which help produce the movement, are called *synergists*. For example, the prime mover for extension of the lower leg is the rectus femoris muscle of the thigh. The synergists for this movement are various other muscles located on the front of the thigh.

Muscles functioning alone cannot produce movements; many other structures must function with them. For instance, as already observed, most muscles bring about movements by pulling on bones across movable joints. But before a skeletal muscle can contract and pull on a bone to move it, the muscle must first be stimulated by nerve impulses. It must also be supplied with oxygen and food; otherwise it soon cannot contract for lack of energy. Energy for all kinds of work done by all kinds of body cells, you will recall, comes from catabolism of foods, which uses oxygen. The respiratory system moves oxygen from the air into the blood. The circulatory system delivers oxygen to body cells. In short then, respiratory, circulatory, nervous, muscular, and skeletal systems all play essential parts in producing normal movements. This fact has practical importance. For example, a person might have perfectly normal muscles and still not be able to move normally. He might have a nervous system disorder that shuts off impulses to certain skeletal muscles and thereby paralyzes them. Poliomyelitis is the great enemy that acts in this way, but so, too, do some other conditions—a hemorrhage in the brain, a brain tumor, or a spinal cord injury, to mention a few. Skeletal system disorders, arthritis especially, can have disabling effects on movement. Muscle functioning, then, depends upon the functioning of many other parts of the body—a fact which illustrates a principle that we shall repeat many times in this book. One way of stating this principle is this: no part of the body lives by itself alone; each part depends upon all other parts for its health and for its very survival.

Contraction of a skeletal muscle does not always produce movement. Sometimes it increases the tension within a muscle but does not decrease the length of the muscle. When the muscle does not shorten, it does not produce any movement. This type of contraction which produces no movement but increases muscle tension is known as *isometric contraction*. The type that produces movement is called *isotonic contraction*. Repeated isometric contractions make muscles grow stronger. Hence the popularizing in recent years of isometric exercises as great muscle builders.

To observe the difference between isometric and isotonic contraction of a muscle, try this simple experiment. Watch the front of your upper arm as you bend your arm at the elbow, moving your hand up toward your shoulder. As you make this movement, the biceps brachii muscle located in the upper arm appears to become shorter and more bulging. Now place one hand on the undersurface of a heavy table or desk top, and, still watching your upper arm, push up on the table top with all your strength. Did your biceps muscle shorten this time as it contracted? Did it produce movement? What kind of contraction did you perform this time—isometric or isotonic?

Types of movements produced by skeletal muscle contractions

Muscles can move some body parts in several directions and others in only two

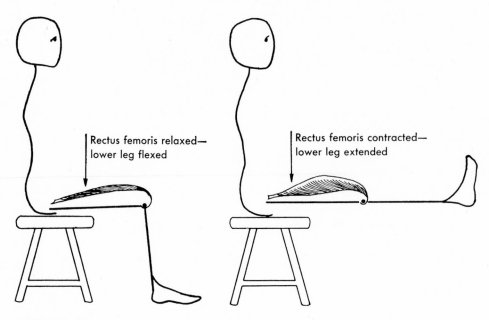

Fig. 3-2

Diagram showing how contraction or shortening of a muscle causes movement. When a muscle contracts, it pulls its insertion bone toward its origin bone. When the rectus femoris muscle contracts, it pulls its insertion bone, the tibia, toward its origin, the hip bone. This straightens the knee joint and extends the lower leg as shown.

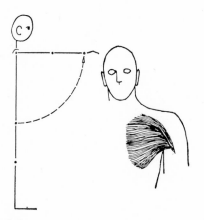

Fig. 3-3

When the pectoralis major muscle (shown on the figure at the right) contracts, it flexes the upper arm at the shoulder joint (figure at the left).

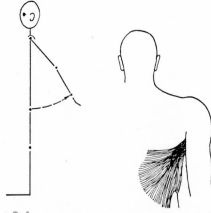

Fig. 3-4

When the latissimus dorsi muscle (shown on the figure at the right) contracts, it extends the upper arm at the shoulder joint (figure at the left).

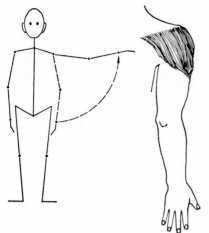

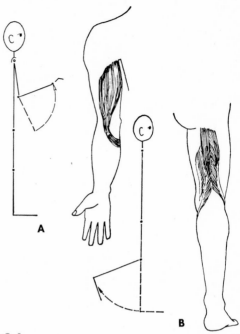

Fig. 3-5

When the deltoid muscle (shown on the figure at the right) contracts, it abducts the upper arm at the shoulder joint (figure at the left).

Fig. 3-6

Flexion of the lower arm and lower leg. **A,** When the biceps brachii muscle (shown at the right) contracts, it flexes the lower arm at the elbow joint (shown at the left). **B,** When the hamstring muscles (shown at the right) contract, they flex the lower leg at the knee joint (shown at the left).

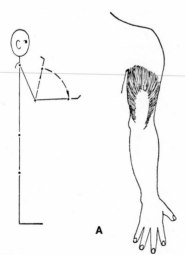

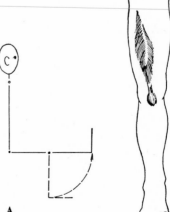

Fig. 3-7

Extension of the lower arm and lower leg. **A,** When the triceps brachii muscle (shown at the right) contracts, it extends the lower arm at the elbow joint (shown at the left). **B,** When the rectus femoris muscle (shown at the right) contracts, it extends the lower leg at the knee joint.

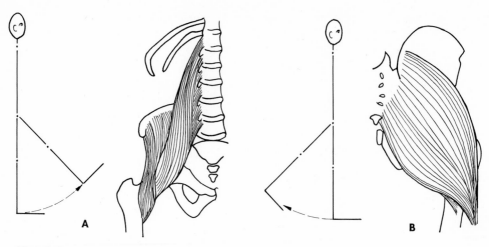

Fig. 3-8

Flexion and extension of the thigh. **A,** When the iliopsoas muscle (shown at the right) contracts, it flexes the thigh at the hip joint (shown at the left). **B,** When the gluteus maximus muscle (shown at the right) contracts, it extends the thigh at the hip joint (shown at the left).

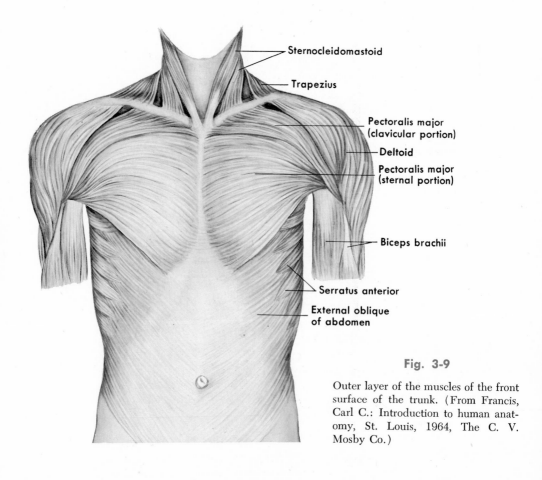

Fig. 3-9

Outer layer of the muscles of the front surface of the trunk. (From Francis, Carl C.: Introduction to human anatomy, St. Louis, 1964, The C. V. Mosby Co.)

directions. As mentioned on page 30, this depends largely on the shapes of the bones at the freely movable joint. Some of the movements we make most often are flexion, extension, abduction, and adduction.

Suppose that you bend one of your arms at the elbow or bend a leg at the knee, or suppose that you bend over forward at the waist or bend your head forward in prayer. With each of these movements, you will have flexed some part of the body—the lower arm, the lower leg, the trunk, and the head, respectively. Most flexions are movements that we commonly describe as bending. The only exception to this rule that I can think of is flexion of the upper arm. Flexing the upper arm means moving it forward from the chest as shown in Figure 3-3. One definition for *flexion* is this: a movement that makes the angle between two bones at their joint smaller than it was at the beginning of the movement.

Extensions are opposite or antagonistic actions to flexions. Thus extensions are movements that make joint angles larger rather than smaller. Extensions are straightening or stretching movements rather than bending movements. Extending the lower arm, for example, is straightening it out at the elbow joint as shown in Figure 3-7, A. Extending the upper arm is stretching it out back from the chest as shown in Figure

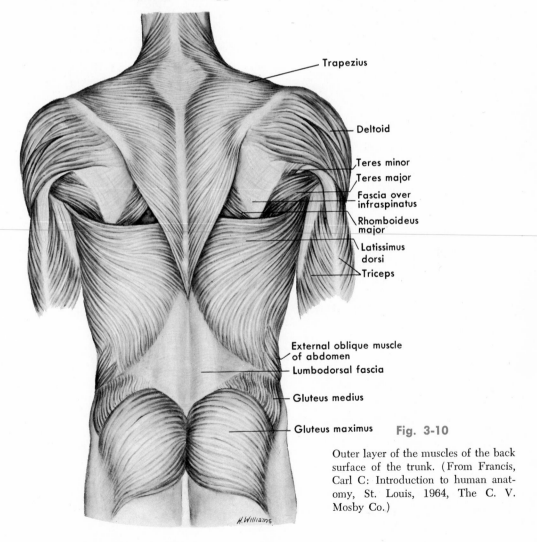

- Trapezius
- Deltoid
- Teres minor
- Teres major
- Fascia over infraspinatus
- Rhomboideus major
- Latissimus dorsi
- Triceps
- External oblique muscle of abdomen
- Lumbodorsal fascia
- Gluteus medius
- Gluteus maximus

H. Williams

Fig. 3-10

Outer layer of the muscles of the back surface of the trunk. (From Francis, Carl C: Introduction to human anatomy, St. Louis, 1964, The C. V. Mosby Co.)

3-4. What movement would you make to extend your lower leg? to extend your thigh or upper leg? Look at Figures 3-7, *B,* and 3-8, *B,* to check your answers. If you straighten your back to stand tall or stretch backwards from your waist, are you flexing or extending your trunk?

Abduction means moving a part away from the midline of the body, such as moving the arms out to the sides.

Adduction means moving a part toward the midline, as bringing the arms down to the sides.

Figures 3-9 to 3-12 show the outer layer

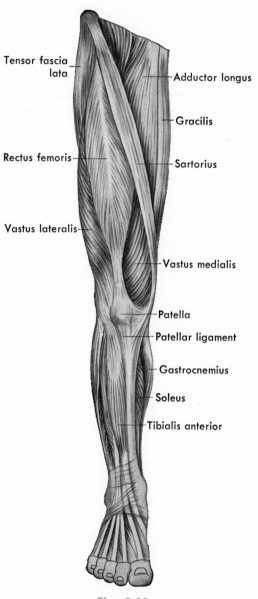

Tensor fascia lata
Adductor longus
Gracilis
Rectus femoris
Sartorius
Vastus lateralis
Vastus medialis
Patella
Patellar ligament
Gastrocnemius
Soleus
Tibialis anterior

Fig. 3-11

Muscles of the front surface of the leg.

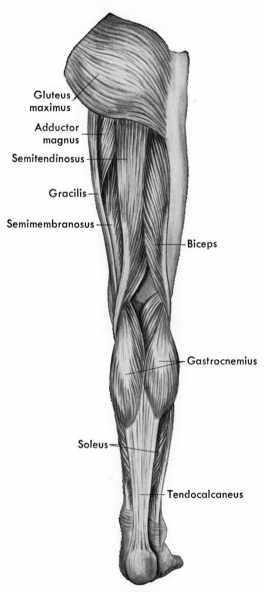

Gluteus maximus
Adductor magnus
Semitendinosus
Gracilis
Semimembranosus
Biceps
Gastrocnemius
Soleus
Tendocalcaneus

Fig. 3-12

Muscles of the back surface of the leg.

Table 3-1. Muscles grouped according to function

Part moved	Flexors	Extensors	Abductors	Adductors
Upper arm	Pectoralis major	Latissimus dorsi	Deltoid	Pectoralis major and latissimus dorsi contracting together
Lower arm	Biceps brachii	Triceps brachii	None	None
Thigh	Iliopsoas	Gluteus maximus	Gluteus medius and minimus	Adductor group
Lower leg	Hamstrings	Quadriceps femoris group	None	None
Foot	Tibialis anterior	Gastrocnemius soleus		
Trunk	Iliopsoas and rectus femoris	Erector spinae (sacrospinalis)		

Table 3-2. Muscle functions, origins, and insertions

Muscle	Function	Insertion	Origin
Pectoralis major	Flexes *upper arm* Helps adduct *upper arm*	Humerus	Sternum Clavicle Upper rib cartilages
Latissimus dorsi	Extends *upper arm* Helps adduct *upper arm*	Humerus	Vertebrae Ilium
Deltoid	Abducts *upper arm*	Humerus	Clavicle Scapula
Biceps brachii	Flexes *lower arm*	Radius	Scapula
Triceps brachii	Extends *lower arm*	Ulna	Scapula Humerus
Iliopsoas	Flexes *trunk*	Ilium Vertebrae	Femur
Iliopsoas	Flexes *thigh*	Femur	Ilium Vertebrae
Gluteus maximus	Extends *thigh*	Femur	Ilium Sacrum Coccyx
Gluteus medius	Abducts *thigh*	Femur	Ilium
Gluteus minimus	Abducts *thigh*	Femur	Ilium
Adductors	Adduct *thigh*	Femur	Pubic bone
Hamstring group	Flexes *lower leg* Helps extend *thigh*	Tibia Fibula	Ischium Femur
Quadriceps femoris group, including rectus femoris	Extends *lower leg* Helps *flex thigh*	Tibia	Ilium Femur

of muscles known as the *superficial muscles. Deep muscles* lie under most of these.

SPECIFIC FACTS ABOUT CERTAIN KEY MUSCLES

Flexors (that is, muscles that flex different parts of the body) produce many of the movements used in walking, sitting, swimming, typing, and a host of other activities. Extensors also function in these activities but perhaps play their most important role in maintaining upright posture. Study Table 3-1 and Figures 3-9 to 3-12 to learn the names of some of the prime flexors, extensors, abductors, and adductors of the body. Consult Table 3-2 to learn their origins and insertions. Keep in mind that muscles move the bone that they have their insertion on. You might use this information in this way—when you learn that a certain muscle is a flexor of the upper arm, you will know immediately that it must insert on what bone? If a muscle is listed as an extensor of the lower leg, what bone or bones must it insert on? Answer review questions 8 and 9 on page 43.

POSTURE

The word "position" can be substituted for the word "posture." Thus, "good posture" means good position of body parts, and this means the position that best favors function. It means the position in which there is a balanced distribution of weight and which, therefore, puts the least strain on muscles, ligaments, and bones. To have good posture in a standing position, for example, you must stand with your head and chest held high, your chin, abdomen, and buttocks pulled in, and your knees bent slightly.

To judge for yourself how important good posture is, consider some of the effects of poor posture. Besides detracting from appearance, poor posture makes a person tire more quickly. It puts an abnormal pull on the ligaments, joints, and bones and therefore often leads to deformities. Poor posture crowds the heart, making it harder for it to contract. Poor posture crowds the lungs, decreasing their breathing capacity.

Skeletal muscles maintain posture by counteracting the pull of gravity. Gravity tends to pull the head and trunk down and forward, but certain back and neck muscles pull just hard enough in the opposite direction to overcome the force of gravity and hold the head and trunk erect. Without continuous muscular pull on the leg bones, for example, your knees would collapse and you would not be able to stand up.

If a person is awake and his muscles are healthy and their nerve supply is intact, the muscles have "good tone," an expression familiar to most of us. It means that the muscles are partly contracted; that is, that only a few fibers in any one muscle are contracted at any one time—enough to make the muscles feel firm to the touch but not enough to cause movement.

MUSCLE DISORDERS

A person with *paralysis* cannot contract his muscles when he wants to; his skeletal muscles do not respond to his will. However, this is not because there is anything wrong with his muscles but because there is disease or injury of his brain or spinal cord or nerves so that they cannot activate his muscles.

Muscle atrophy is muscle shrinkage, a decrease in muscle size due to disuse. Thus many conditions can cause atrophy—such as having a leg in a cast, being a bed patient for a long time, or paralysis.

Muscle hypertrophy is the opposite of atrophy; that is, it is an increase in size resulting from increased use. Skeletal muscles may hypertrophy as a result of exercise. The heart frequently hypertrophies from overwork.

Outline summary—The muscular system

MUSCLE TISSUE

1. Striated muscles; also called skeletal muscle or voluntary muscle
2. Branching muscle; also called cardiac muscle
3. Nonstriated muscle; also called smooth muscle, visceral muscle, or involuntary muscle

GENERAL FACTS ABOUT SKELETAL MUSCLES

1. Functions—movements and posture
2. Parts of a skeletal muscle
   a. Origin—attachment to relatively immovable bone
   b. Insertion—attachment to bone that moves
   c. Body—main part of muscle
3. How muscles produce movement—as muscle contracts, it pulls insertion bone nearer origin bone; movement occurs at joint between origin and insertion
4. Groups of muscles usually contract to produce a single movement
   a. Prime mover—the muscle whose contraction is mainly responsible for producing a given movement
   b. Synergists—muscles whose contractions help produce a given movement
5. Normal muscle functioning depends upon normal functioning of various other structures, notably nerves and joints
6. Not all muscle contractions produce movements; isometric contractions increase the tension in muscles without producing movements; isotonic contractions produce movements

Types of movements produced by skeletal muscle contractions

1. Flexion—making angle at joint smaller
2. Extension—making angle at joint larger
3. Abduction—moving a part away from midline
4. Adduction—moving a part toward midline

SPECIFIC FACTS ABOUT CERTAIN KEY MUSCLES

See Tables 3-1 and 3-2

POSTURE

1. Posture means position of body parts
2. Good posture important for many reasons

—for example, to prevent fatigue and bone and joint deformities
3. Skeletal muscles maintain posture by counteracting the pull of gravity—by maintaining tone (partial contraction)

MUSCLE DISORDERS

1. Paralysis—loss of ability to produce voluntary movements
2. Atrophy—decrease in muscle size
3. Hypertrophy—increase in muscle size

Review questions—The muscular system

1. Compare the three kinds of muscle tissue as to location, microscopic appearance, and nerve control.
2. Explain why skeletal muscle functions are so important. What are the general functions of skeletal muscle?
3. Explain the terms flexion, extension, abduction, and adduction. Give an example of each.
4. Explain how skeletal muscles, bones, and joints work together to produce movements.
5. Why can a spinal cord injury be followed by muscle paralysis?
6. Can a muscle contract very long if its blood supply is shut off? Give a reason for your answer.
7. The correct term to substitute in the expression "bending your knee" is (extending? flexing?) your lower leg.
8. What is the name of the main muscle which
   a. Flexes the upper arm?
   b. Flexes the lower arm?
   c. Flexes the thigh?
   d. Flexes the lower leg?
   e. Extends the upper arm?
   f. Extends the lower arm?
   g. Extends the thigh?
   h. Extends the lower leg?
9. Give the approximate location of each of the following muscles and tell what movement it produces:
   biceps brachii   pectoralis major
   hamstrings   quadriceps femoris group
   deltoid   latissimus dorsi
10. Explain the following terms:
   posture   hypertrophy
   muscle tone   muscle tone
   atrophy   good posture

How the nervous system controls body functions

IV

The nervous system

Any group of individuals who work together on anything have to have a boss. If their many separate jobs together are going to accomplish some big complex task, then obviously what each worker does must be controlled or regulated. This principle holds true regardless of the nature of the complex task to be done. Suppose the task is that of giving good care to hospitalized sick people. Can you imagine what would happen if there were no way of letting each individual worker know just what he was to do as his part of the job of caring for the hospital's patients? Pretty clearly, chaos would result. Some system for communicating with individual workers and controlling them is absolutely essential in any large organization.

The normal body accomplishes a gigantic and enormously complex job—that of keeping itself alive and healthy. Each one of its billions of individual cells performs some function that is a part of this big function. Control of the body's billions of cells is accomplished largely by two communication systems, namely the nervous system and the endocrine system. Both systems transmit information from one part of the body to another, but they do it in different ways. The nervous system transmits information by means of nerve impulses conducted by nerve cells from one structure to another. The endocrine system does it by means of chemicals secreted by ductless glands into the blood stream and circulated from the glands to other parts of the body. Communication makes possible control. Nerve impulses and hormones communicate information to body structures which controls their activities, increasing them or decreasing them as needed for healthy survival. In other words, the communication systems of the body are also its control and integrating systems. By integrating systems

47

we mean devices that weld the body's hundreds of different functions into its one overall function—that of keeping itself alive and healthy. We shall discuss the nervous system in this chapter and the endocrine system in Chapter 10. Our plan for this chapter is to relate basic information about the structure and function of the cells of the nervous system and then of its organs and their coverings.

CELLS OF THE NERVOUS SYSTEM

Cells that specialize in transmitting impulses are called *neurons* or *nerve cells.* Organs of the nervous system consist chiefly of neurons. A special type of connective tissue cell known as *neuroglia* also occurs in abundance in the nervous system. One reason for mentioning neuroglia is that one of the commonest types of brain tumor—called the *glioma*—develops from it. Neuroglia vary in size and shape. Some are relatively large cells shaped somewhat like stars because of the many threadlike extensions that jut out from their surfaces. (These neuroglia are called *astrocytes,* a word which means star-shaped cells.) The threadlike branches of astrocytes attach them to both neurons and small blood vessels, holding these structures close to each other. Some neuroglia, called *microglia* because of their small size, move about in brain tissue if it becomes inflamed. They act as scavengers, engulfing and destroying microorganisms and cellular debris by a process called *phagocytosis.*

Each neuron consists of three main parts: a main part called the *neuron cell body,* one or more branching projections called *dendrites,* and one elongated projection known as an *axon.* By definition, dendrites are the processes (projections) that transmit impulses to the neuron cell bodies and axons are the processes that transmit impulses away from them.

Classified according to function, there are three types of neurons: sensory neurons, motoneurons, and interneurons. *Sen-*

sory neurons transmit impulses to the spinal cord and the brain. *Motoneurons* transmit impulses away from the brain and away from the cord out to muscles and glands. *Interneurons* conduct impulses from sensory neurons to motoneurons. Sensory neurons are also called afferent neurons; motoneurons are called efferent neurons; interneurons are called internuncial, intercalated, central, or connecting neurons.

Conduction of nerve impulses

Nerve impulses are conducted over what we shall call "neuron pathways." The simplest kind of neuron pathway consists of two types of neurons, namely sensory neurons and motoneurons, conducting in that order. Impulse conduction starts when an adequate stimulus of some kind acts on the endings of the dendrites of sensory neurons. Because these structures receive stimuli, they are called *receptors.* Receptors are usually defined as the endings of sensory neuron dendrites, but since impulse conduction begins in receptors, it seems more logical to me to define them as the beginnings (instead of the endings) of sensory dendrites. From receptors, impulses travel the full length of the sensory neuron's dendrite, cell body, and axon to the branching brushlike ends of the axon. In the simplest neuron pathways, the ends of sensory neuron axons contact the dendrites or the cell bodies of motoneurons. *Synapse* is the name for these places of contact between the axon endings of one neuron and the dendrites or cell body of another neuron. From the synapse, nerve impulses are conducted by motoneuron dendrites, cell bodies, and axons. Axons of somatic motoneurons—that is, of neurons whose cell bodies lie in the anterior gray columns of the spinal cord—leave the cord by way of the anterior roots of spinal nerves (Figure 4-1) and extend in spinal nerves out to skeletal muscles. Impulses conducted by somatic motoneurons stimulate skeletal

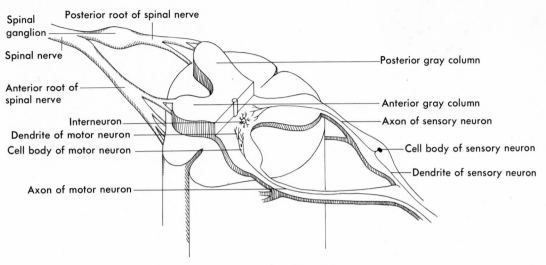

Spinal ganglion

Posterior root of spinal nerve

Spinal nerve

Anterior root of spinal nerve

Interneuron

Dendrite of motor neuron

Cell body of motor neuron

Axon of motor neuron

Posterior gray column

Anterior gray column

Axon of sensory neuron

Cell body of sensory neuron

Dendrite of sensory neuron

Fig. 4-1

Cross section of the spinal cord. The left side of diagram shows macroscopic structures only. The right side of diagram shows locations of the dendrite, cell body, and axon (all microscopic structures) of a sensory neuron, an interneuron, and a motoneuron, the structures which compose a three-neuron reflex arc.

muscles to contract. Skeletal muscles might, in other words, be said to put into effect the information brought to them by somatic motoneurons. For this reason, skeletal muscles are called *somatic effectors*.

The neuron pathway described in the preceding paragraph is called a *reflex arc*. Recall that it consists of sensory neurons synapsing with motoneurons. Because these simplest reflex arcs consist of only two kinds of neurons, they are known as *two-neuron arcs*. A slightly more complex type of reflex arc consists of sensory neurons that synapse with interneurons which synapse with motoneurons—*three-neuron arcs*, these are called. Many more complex reflex arcs also exist but cannot be described adequately in this brief book.

If you examine Figure 4-1 carefully, you can learn a number of important facts from it. Dendrites, cell bodies, and axons of neurons, which are microscopic structures, are located in spinal nerves and roots and in the spinal cord, which are macroscopic structures. Dendrites of sensory neurons, for example, are located in the spinal nerves and many of their branches. Sensory dendrites are necessarily long processes because they extend from the skin

and other outlying parts of the body all the way to the spinal column. Cell bodies of sensory neurons are located in rounded structures called ganglia which are found on the posterior roots of spinal nerves. Each of these posterior root ganglia contains hundreds of sensory neuron cell bodies. The axons of sensory neurons are located in the posterior roots of the spinal nerves. These axons are short and terminate in the gray matter of the spinal cord where they synapse with interneurons (or in two-neuron arcs with motoneurons). In spinal cord reflex arcs, interneurons lie wholly within the gray matter of the spinal cord as you can see in Figure 4-1.

Dendrites and cell bodies of motoneurons whose axons terminate in skeletal muscle are located in the anterior horns of the gray matter of the spinal cord—a fact of great practical importance. For instance, it explains why poliomyelitis causes paralysis. The polio virus attacks the anterior gray horns, injuring and destroying motoneuron cells, thereby making them unable to conduct impulses to the skeletal muscles which they supply.

The term *"reflex center"* means the center of a reflex arc. The first part of a reflex arc consists, as we have seen, of sensory neurons and the last part consists of motoneurons. In spinal cord arcs, the sensory neurons conduct impulses to the cord and motoneurons conduct them away from the cord out to muscles. The reflex centers, then, of all spinal cord arcs lies in spinal cord gray matter. It consists of interneurons, as shown in Figure 4-1, or, in two-neuron arcs, it is simply the synapses between sensory and motoneurons. A broad definition of reflex center is that it is the place in a reflex arc where incoming impulses become outgoing impulses.

When nerve impulses have traveled a complete reflex arc—that is, all the way from receptors to effectors (muscles or glands)—they stimulate these structures to respond. Their response is called a *reflex.* The cause of impulse conduction by a reflex arc is stimulation of receptors; the result of impulse conduction by a reflex arc is a response by effectors, a reflex—for example, a skeletal muscle contraction.

BRAIN AND CORD COVERINGS AND FLUID SPACES

Nerve tissue is not one of our sturdy tissues. Even moderate pressure can kill nerve cells, so nature safeguards the organs that consist chiefly of nerve cells (namely the brain and spinal cord) by surrounding them with fluid, membranes, and bones. Cranial bones encase the brain; vetebrae encase the spinal cord. A tough membrane, the *meninges,* lines cranial bones and vertebrae. Three layers of tissue form the meninges; they are the dura mater, arachnoid membrane, and pia mater. The middle layer of the meninges, the arachnoid membrane, resembles a cobweb with fluid filling in its spaces. (The word "arachnoid" means cobweb-like. It comes from Arachne, the name of the girl who was changed into a spider because she boasted of the fineness of her weaving. At least so an ancient Greek myth tells us.)

Meningitis is an inflammation of the meninges. In this disease, cerebrospinal fluid accumulates in the arachnoid spaces and puts added pressure on the brain and cord. Because this has an injurious effect on these delicate structures, the physician may do a *lumbar puncture* to relieve the pressure. He inserts a needle between two of the lumbar vertebrae into the arachnoid space and withdraws some of the fluid. Through what layers of the meninges does he have to push his needle in order to enter the arachnoid space? Find your answer by looking at Figure 4-2.

Cerebrospinal fluid is present not only in the arachnoid spaces around the brain and spinal cord but also in spaces inside the brain called *ventricles* and in the small tubelike space (the *central canal*) inside the cord. Cerebrospinal fluid continually forms from blood in capillaries in the brain ventricles and continually returns to blood in veins on the outside of the brain. Sometimes a brain tumor or other abnormality blocks the return of cerebrospinal fluid to the blood. Then fluid accumulates in the ventricles or arachnoid spaces of the brain This condition is called *hydrocephalus,* or in popular language, "water on the brain."

THE ORGANS OF THE NERVOUS SYSTEM

The organs of the nervous system are the brain, the spinal cord, and the numerous nerves of the body. Often the brain and spinal cord together are referred to as the *central nervous system* (CNS)—an appropriate name for them in view of their central location in the body and their central role in the functioning of the nervous system. In contrast, all the nerves of the

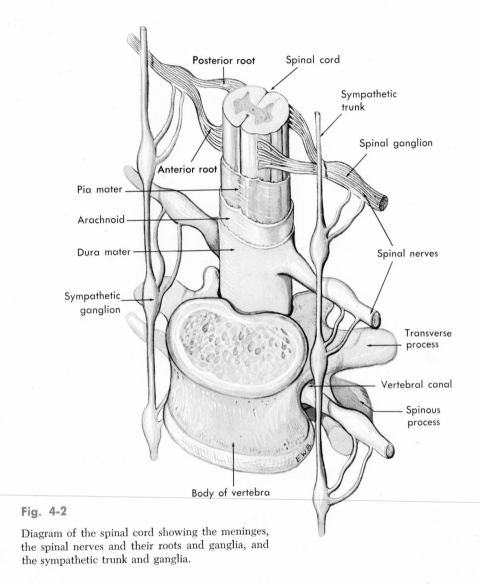

Fig. 4-2

Diagram of the spinal cord showing the meninges,
the spinal nerves and their roots and ganglia, and
the sympathetic trunk and ganglia.

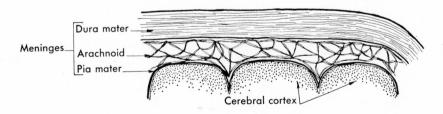

Fig. 4-3

Schematic drawing showing the structure of the
meninges around the brain.

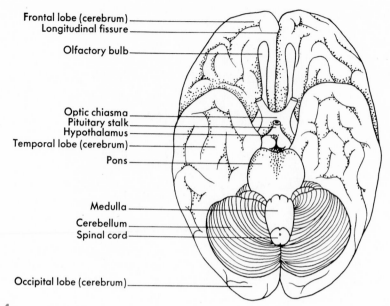

Frontal lobe (cerebrum)
Longitudinal fissure

Olfactory bulb

Optic chiasma
Pituitary stalk
Hypothalamus
Temporal lobe (cerebrum)
Pons

Medulla
Cerebellum
Spinal cord

Occipital lobe (cerebrum)

Fig. 4-4

Undersurface of the human brain. Note the location of the hypothalamus and of the stalk of the pituitary gland in the middle area.

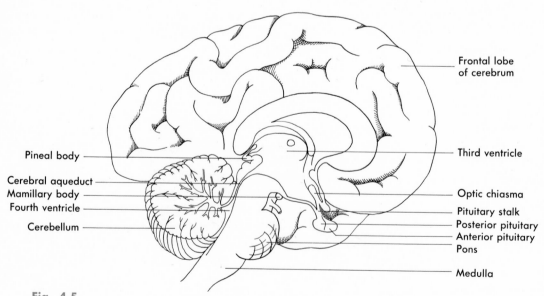

Frontal lobe
of cerebrum

Pineal body

Third ventricle

Cerebral aqueduct
Mamillary body
Fourth ventricle
Cerebellum

Optic chiasma

Pituitary stalk
Posterior pituitary
Anterior pituitary
Pons

Medulla

Fig. 4-5

Longitudinal section of the brain.

body together are referred to as the *peripheral nervous system* (PNS). Peripheral means outlying. So nerves, reaching out as they do from brain and cord to all parts of the body, seem well named as the peripheral nervous system.

Several other terms are also used frequently in connection with the nervous system: the voluntary nervous system, the involuntary nervous system, the autonomic nervous system, the sympathetic system, and the parasympathetic system. Later on in this chapter we shall define these terms.

The brain

The brain consists of several parts, the most prominent of which are the cerebrum, the cerebellum, the midbrain, the pons, and the medulla. (Together, the midbrain, pons, and medulla seem to form a stem for the rest of the brain. In fact, they are often designated jointly as the *brainstem*.)

Two less prominently located parts of the brain—the thalamus and the hypothalamus—have captured a great deal of attention in recent years. They are the largest of several structures which together make up the *diencephalon,* the area of the brain tucked in between the cerebrum and the brainstem.

The cerebrum. The largest part of the brain is the cerebrum (Figures 4-4 and 4-5). A deep groove, the longitudinal fissure, divides it into two hemispheres connected only in their lower middle portion. The outer layer of the cerebral hemispheres is called the *cerebral cortex.* (The word cortex means bark or outer layer.) The cerebral cortex is not smooth; rather, it has many rounded ridges *(convolutions)* with grooves between them. Because of its pinkish-gray color the cerebral cortex is said to be made up of *gray matter.* Most of the interior of the cerebrum, on the other hand, consists of white matter. The different areas of the cerebral cortex are called "lobes." The frontal, parietal, temporal, and occipital lobes are the main ones. As you

might guess, the two frontal lobes lie under the frontal bones, the parietal lobes lie under the parietal bones, and so on.

What functions does the cerebrum perform? This is a hard question to answer briefly because the neurons of the cerebrum do not function alone. They function with many other neurons located in many other parts of the brain and in the spinal cord. Neurons of these structures are continually bringing impulses to cerebral neurons and continually transmitting impulses away from them. If all these other neurons were functioning normally and only cerebral neurons were not functioning here are some of the things that you could not do. You could not think or will. You could not decide to make the simplest movement and make it. You could not see or hear. You would not experience any of the other sensations that make life so rich and varied. Nothing would anger you or frighten you. Nothing would bring you great joy or great sorrow. In fact, you would be unconscious. The following five

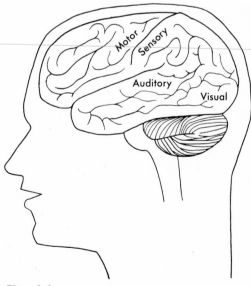

Fig. 4-6

Side view of the brain, showing some identified functional areas of the cerebral cortex.

terms, then, summarize cerebral functions—consciousness, mental processes, sensations, emotions, and voluntary movements. Figure 4-6 shows the main functions of a few areas of the cerebral cortex.

The cerebellum. The cerebellum is the second largest part of the human brain. It lies under the occipital lobe of the cerebrum. Like the cerebrum, the cerebellum has an outer layer of gray matter and an interior of almost all white matter.

Most of our knowledge about cerebellar functions has come from observing patients who have some sort of disease of the cerebellum and animals who have had the cerebellum removed. From such observations we know that the cerebellum plays an essential part in the production of normal movements. Perhaps a few examples will make this clearer. A patient who has a tumor of the cerebellum frequently loses his balance and topples over. He may reel like a drunken man when he walks. He probably cannot coordinate his muscles normally. He may complain, for instance, that he is clumsy about everything that he does—that he cannot even pound a nail or draw a halfway straight line. With the loss of normal cerebellar functioning, he has lost the ability to make precise movements. To summarize, the general function of the cerebellum is to so stimulate and inhibit groups of muscles as to produce smooth coordinated movements, maintain equilibrium, and sustain normal postures.

The medulla. The lowest part of the brain and the only division of the brainstem that we shall discuss is the medulla. It is an enlarged extension of the spinal cord, so it lies just above the large opening (foramen magnum) in the occipital bone. (The spinal cord starts just below this opening.) The medulla claims our attention as the most vital part of the entire nervous system. It contains the vital centers—namely the cardiac centers, the vasomotor centers, and the respiratory centers. Each

center consists of groups of neurons. Cardiac center neurons relay impulses to the heart, controlling the rate and the strength of the heart beat. Vasomotor center neurons relay impulses to blood vessels, controlling their diameter, that is, either constricting or dilating them. Respiratory center neurons control respiratory muscle contractions, thereby controlling the rate and depth of respirations. If neurons of the medulla's vital centers fail to function (perhaps because they receive too little oxygen or because they receive too much pressure from a brain tumor or skull injury), death quickly follows.

The thalamus

The term "thalamus" is singular, suggesting the presence of one thalamus, but actually we have two of these organs, a right thalamus and a left thalamus. The right thalamus consists of a rounded mass of gray matter located deep inside the right half of the cerebrum; the left thalamus is a similar mass inside the left cerebral hemisphere. Microscopically, the thalamus consists chiefly of dendrites and cell bodies of neurons whose axons extend to various sensory areas of the cerebral cortex.

The thalamus functions to help produce sensations. Its neurons constitute one part of the neuron pathways which transmit the impulses that result in sensations. All sensory impulses (except perhaps those responsible for the sense of smell) are conducted to the thalamus by neurons which synapse with neurons whose dendrites and cell bodies lie in the thalamus and whose axons extend to a sensory area in the cerebral cortex. Sensory impulses are conducted to the thalamus from below and its neurons relay them to the cerebral cortex. In other words, the thalamus functions as a relay station for sensory impulses.

The thalamus almost surely is the part of the brain which associates sensations with emotions. Each of us experiences each sensation with some degree of pleasantness or unpleasantness. Some sensations we find extremely pleasant, some extremely unpleasant and many others—perhaps the majority—we have very little feeling about either one way or the other. For some reason, these feelings seem to come from the arrival of sensory impulses in the thalamus.

The hypothalamus

Part of the hypothalamus, as the name suggests, lies under each thalamus. Another part of it lies above the midbrain. The stalk and posterior part of the pituitary gland, for example, are part of the hypothalamus, and as you can see in Figure 4-5, they are located just above and in front of the midbrain.

The old adage, "Don't judge by appearances," applies well to an appraisal of the importance of the hypothalamus. Measured by size, the hypothalamus is one of the least significant parts of the brain, but measured by its contribution to healthy survival, the hypothalamus is an extremely important structure.

The hypothalamus acts as the major center for controlling the autonomic nervous system and therefore helps control most of our internal organs. It controls such things as the heart's beating, the constriction or dilatation of blood vessels, the contractions of the intestine, and the emptying of the bladder. The hypothalamus controls hormone secretion by both the anterior and posterior pituitary glands. Indirectly, therefore, it also helps control the amount of hormones secreted by practically all of the other endocrine glands. The hypothalamus acts as the center for controlling appetite and for this reason plays an important part in regulating how much food an individual eats and how much he weighs. The hypothalamus functions in some way to keep us awake. Evidence of this is the fact that a person with disease or injury of the hypothalamus sleeps continually and can hardly be awakened.

Last but not least, the hypothalamus probably contains reward and punishment centers (also called pleasure and displeasure centers), and so functions to produce some of our basic urges or drives. Experiments performed on rats and other animals in recent years have revealed some exciting information about this. For example, rats pressed levers hundreds and even thousands of times an hour when these levers were arranged so as to stimulate tiny electrodes implanted at certain parts in the hypothalamus. That these animals must have found this stimulation pleasant seems an understatement. In fact, even when they were hungry, when they had not had food for many hours, they would still go on pressing the lever rather than stop to eat food when it was set before them. Other experiments seem to indicate that the hypothalamus also contains punishment or displeasure centers and that both reward and punishment centers exist in several other parts of the brain as well.

The spinal cord

If you are of average height, your spinal cord is about 17 or 18 inches long. It lies inside the spinal column in the spinal cavity and reaches from the occipital bone down to the bottom of the first lumbar vertebrae. (Place your hand on your hips and they will line up with your fourth lumbar vertebrae; your spinal cord ends shortly above this level.) The spinal cord meninges, however, continue on down almost to the end of the spinal column—a convenient fact for a physician to know because it means that he can do a lumbar puncture without fear of damaging the cord.

The spinal cord carries on two important functions. It conducts impulses between the brain and the many other parts of the body. This is a two-way conduction, that is, upward toward the brain by sensory nerve fibers and downward from it by motor fibers. Therefore if the spinal cord is injured, it may mean that impulses can

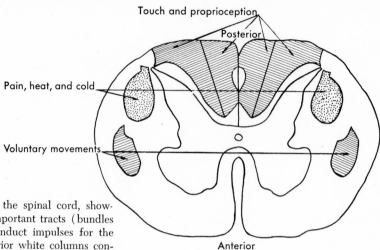

Touch and proprioception

Posterior

Pain, heat, and cold

Voluntary movements

Anterior

Fig. 4-7

Cross-sectional diagram of the spinal cord, showing the location of some important tracts (bundles
of nerve fibers), which conduct impulses for the
functions listed. The posterior white columns conduct impulses for touch and proprioception; the
lateral spinothalamic tracts, those for pain, heat,
and cold; and the lateral corticospinal tracts, those
for voluntary movements.

no longer travel up to the brain from parts
located below the injury or come from
the brain down to these parts. In other
words, a spinal cord injury can produce
both a loss of sensation (anesthesia) and a
loss of the ability to make voluntary movements (paralysis).

The spinal cord's other function is to
serve as a center for reflexes. Best known
of these perhaps is the knee jerk reflex.
The many reflex centers in the cord function to automatically produce many of our
movements without making it necessary for
us to consciously will every move that we
make. The meaning of the term "reflex center" is explained on page 50.

In summary, the spinal cord performs
two important functions: it serves as the
great conduction pathway between the
brain and other parts of the body, and it
serves as the reflex center for many different reflexes.

The cranial nerves

Twelve pairs of cranial nerves attach to
the undersurface of the brain. Their fibers
conduct impulses between the brain and
various structures in the head and neck
and in the thoracic and abdominal cavities.
For instance, the first cranial nerve (the
olfactory nerve) conducts impulses from
the nose to the brain where these impulses
produce the sensation of smell. The third
cranial nerve (oculomotor nerve), on the
other hand, conducts impulses from the
brain to certain muscles of the eye, where
these impulses cause contractions that
move the eye. The names of each of the
cranial nerves and a brief description of
their functions are listed in Table 4-1.

The spinal nerves

Thirty-one pairs of nerves attach to the
spinal cord. They do not have special
names as cranial nerves do but are num-
bered according to the segment of the cord
that they attach to. For example, we might
speak of spinal nerve "C-1," meaning the
pair of nerves attached to the first segment
of the cervical part of the cord, or spinal
nerve "T-8," meaning the pair attached
to the eighth segment of the thoracic part
of the cord, and so on. There are eight
pairs of cervical, twelve pairs of thoracic,
five pairs of lumbar, five pairs of sacro-
spinal, and one pair of coccygeal nerves.

Table 4-1. Cranial nerves

Nerve*	Conducts impulses	Functions
I. Olfactory	From nose to brain	Sense of smell
II. Optic	From eye to brain	Vision
III. Oculomotor	From brain to eye muscles	Eye movements
IV. Trochlear	From brain to external eye muscles	Eye movements
V. Trigeminal (or trifacial)	From skin and mucous membrane of head and from teeth to brain; also from brain to chewing muscles	Sensations of face, scalp, and teeth; chewing movements
VI. Abducens	From brain to external eye muscles	Turning eyes outward
VII. Facial	From taste buds of tongue to brain; from brain to face muscles	Sense of taste; contraction of muscles of facial expression
VIII. Auditory (or acoustic)	From ear to brain	Hearing; sense of balance
IX. Glossopharyngeal	From throat and taste buds of tongue to brain; also from brain to throat muscles and salivary glands	Sensations of throat; taste; swallowing movements; secretion of saliva
X. Vagus	From throat, larynx, and organs in thoracic and abdominal cavities to brain; also from brain to muscles of throat and to organs in thoracic and abdominal cavities	Sensations of throat, larynx, and of thoracic and abdominal organs; swallowing, voice production, slowing of heartbeat, acceleration of peristalsis
XI. Spinal accessory	From brain to certain shoulder and neck muscles	Shoulder movements; turning movements of head
XII. Hypoglossal	From brain to muscles of tongue	Tongue movements

*The first letters of the words of the following sentence are the first letters of the names of cranial nerves: "On Old Olympus' Tiny Tops A Finn and German Viewed Some Hops." Many generations of students have used this or a similar sentence to help them remember the names of cranial nerves.

Spinal nerves conduct impulses between the spinal cord and parts of the body not supplied by cranial nerves. As you can see in Figure 4-1, all spinal nerves contain both sensory and motor fibers. Spinal nerves therefore function to make possible both sensations and movements. This means that if a patient has a disease or injury that damages a spinal nerve, he will not feel anything in the part of the body that nerve supplies nor will he be able to move this part. Disease or injury of another part of the nervous system may also cause loss of sensation and voluntary movement. Do you recall what structure this is? (The answer is on page 56.)

THE AUTONOMIC OR INVOLUNTARY NERVOUS SYSTEM

The term "autonomic nervous system" is somewhat misleading. It suggests a separate and independent system of the body, but this is not so. The autonomic nervous system is a part of the body's one big complex nervous system. Certain ganglia connected by nerve fibers to the spinal cord and brainstem and certain nerves connected with the ganglia constitute the organs of the autonomic nervous system. Neurons of the autonomic nervous system are those that conduct impulses out from the brainstem or spinal cord to smooth muscle, cardiac muscle, and glandular epithelial tissue. Autonomic neurons are motoneurons—not, however, somatic motoneurons, that conduct impulses to skeletal muscles or somatic effectors, but visceral motoneurons that conduct impulses to smooth muscle, cardiac muscle, and glands or visceral effectors. Note that these are structures that we cannot control voluntarily. Except for a few rare individuals, we cannot, for example, will the smooth muscle in our intestines

to contract more rapidly to increase peristalsis, our heartbeats to speed up or to slow down, or our glands to increase or decrease their secretions. One way to define the autonomic nervous system, then, is as the part of the nervous system that controls the involuntary or automatic functions of the body, the processes most responsible for maintaining life.

The autonomic nervous system has two divisions, known as the *sympathetic nervous system* and the *parasympathetic nervous system*. When impulses over sympathetic fibers control automatic functions, rapid changes take place within our bodies. Table 4-2 indicates some of these changes. The heart beats faster, most blood vessels constrict, causing blood pressure to shoot up, blood vessels in the skeletal muscles dilate to furnish them with more blood, sweat glands and adrenals secrete more abundantly, salivary and other digestive glands secrete more sparingly, and peristalsis becomes sluggish. These changes get us ready for strenuous muscular work or, as someone has so vividly stated, "sympa-

Table 4-2. Autonomic functions

Structure	Parasympathetic control	Sympathetic control
Heart muscle	Slower heartbeat	Faster heartbeat
Smooth muscle of most blood vessels	None	Constricted
Smooth muscle of blood vessels in skeletal muscles	None	Dilated
Smooth muscle of digestive tract	Increased peristalsis; defecation	Decreased peristalsis; inhibition of defecation
Adrenal medulla	Decreased epinephrine secretion	Increased epinephrine secretion
Sweat glands	None	Increased sweat secretion
Digestive glands	Increased secretion of digestive juices	Decreased secretion of digestive juices

thetic impulses prepare us for flight or fight." Therefore the sympathetic nervous system can be thought of as the emergency nervous system. It takes over control of many of our internal functions when we are angry, frightened, worried, or when we are exercising hard—in short, whenever we are undergoing any kind of stress. If you observe patients before surgery or observe yourself, for that matter, when you are terribly frightened or have been running fast, you will probably notice some of the typical effects of sympathetic functioning just mentioned. Under ordinary circumstances when we are not struggling with mental or physical trials, parasympathetic control of most of our automatic functions dominates.

Later on we shall learn the name of a hormone that produces changes similar to those produced by the sympathetic system. (If you think you already know the name of this hormone, you can check your answer on page 136.)

Study Table 4-2 to learn how certain body functions differ according to whether sympathetic or parasympathetic impulses dominate their control.

THE SENSE ORGANS

If you were asked to name the sense organs, what organs would you name? Can you think of any besides the eyes, ears, nose, and taste buds? Actually there are millions of other sense organs—all receptors are microscopic-size sense organs. Receptors, you will recall, are the beginnings of dendrites of sensory neurons.

Receptors are scattered about generously in almost every part of the body. To demonstrate this fact, try pricking any point of your skin with a fine needle. You can hardly miss stimulating at least one receptor and almost instantaneously experiencing a sensation of mild pain. Stimulation of some receptors leads to the sensation of heat. Stimulation of other receptors gives the sensation of cold, and stimulation of

still others give the sensation of touch or pressure. When special receptors in the muscles and joints are stimulated, you sense the position of the different parts of the body and know whether they are moving or not and in which direction they are moving without even looking at them. Perhaps you have never realized that you have this sense of position and movement— a sense called *proprioception* or *kinesthesia.* Let us turn our attention now to two complex and remarkable sense organs—the eyes and ears.

The eye

When you look at a person's eye, you see only a small part of the whole eye. Three layers of tissue form the eyeball: the sclera, the chorioid, and the retina. The outer layer or *sclera* consists of tough fibrous tissue. What we call the "white" of the eye is part of the front surface of the sclera. The other part of the front surface of the sclera is sometimes spoken of as the "window" of the eye because of its transparency. At a casual glance, however, it does not look transparent but appears blue or brown or gray or green because it lies over the iris, the colored part of the eye. The name of the transparent part of the sclera is the *cornea.* Mucous membrane known as the *conjunctiva* covers the entire front surface of the eyeball and lines both the upper and lower lids.

Two involuntary muscles make up the front part of the middle or *chorioid* coat of the eyeball. One is the *iris,* the colored structure seen through the cornea, and the other is the ciliary muscle. (See Figure 4-9.) What appears to be a black center in the iris is really a hole in this doughnut-shaped muscle; it is the *pupil* of the eye. Some of the fibers of the iris are arranged like spokes in a wheel. When they contract, the pupils dilate, letting in more light rays. Other fibers are circular. When they contract, as they do in bright light, the pupils constrict, letting in fewer light rays.

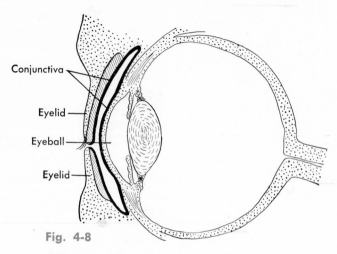

Fig. 4-8

Cross-sectional diagram of the eye.

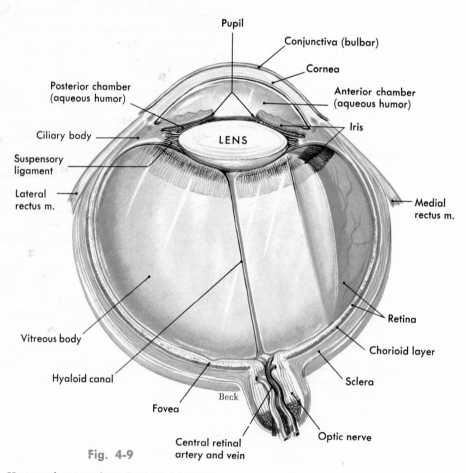

Fig. 4-9

Horizontal section throught the left eyeball.

The *lens* of the eye lies directly behind the pupil. It is held in place by a ligament attached to the ciliary muscle. When the ciliary muscle contracts, the shape of the lens changes. It becomes more bulging and curved, and this change brings near objects into focus. Far objects, on the other hand, come into focus when the ciliary muscle is relaxed and the lens has a flatter, less curved shape.

The *retina* or innermost coat of the eyeball contains microscopic structures called rods and cones because of their shapes. Both are receptors for vision. Dim light can stimulate the rods, but fairly bright light is necessary to stimulate the cones. In other words, *rods* are the receptors for night vision and *cones* for daytime vision. (Cones are also the receptors for color vision.)

Fluids fill the hollow inside of the eyeball. They maintain the normal shape of the eyeball and help refract light rays, that is, bend them so as to bring them to a focus on the retina. *Aqueous humor* is the name of the fluid in front of the lens (in the *anterior cavity* of the eye) and *vitreous humor* is the name of the jelly-like fluid behind the lens (in the *posterior cavity*).

The ear

The ear is much more than a mere appendage on the side of the head. A large part of the ear—and by far its most important part—lies hidden from view deep inside the temporal bone. Part of the external ear and all of the middle ear and the internal ear are located here.

The *external ear* has two parts; the pinna (or auricle) and the ear canal (or external acoustic meatus). The *pinna* is the appendage on the side of the head. The *ear canal* is a curving tube in the temporal bone that leads from the pinna to the middle ear.

The *middle ear* (or tympanic cavity) is a tiny cavity hollowed out of the temporal bone. This cavity is lined by mucous membrane and contains three very small bones. The names of these ear bones are Latin words that describe their shapes—*malleus* (hammer), *incus* (anvil), and *stapes* (stirrup). The *tympanic membrane* (commonly called the eardrum) separates the middle ear from the external ear canal. The "handle" of the malleus attaches to the inside of the tympanic membrane, and the "head" attaches to the incus. The incus attaches to the stapes, and the stapes fits into a small opening, the *oval window*, that opens into the internal ear. A point worth mentioning because it explains the frequent spread of infection from the throat to the ear is the fact that a tube *(eustachian tube)* connects the throat with each middle ear. The mucous lining of both middle

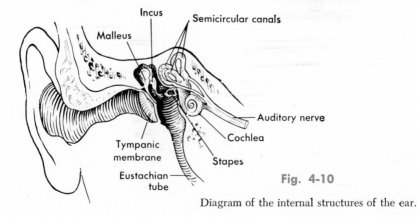

Fig. 4-10

Diagram of the internal structures of the ear.

ears, eustachian tubes, and the throat are extensions of one continuous membrane. Consequently a sore throat may spread to produce a middle ear infection (*otitis media*) as well. Moreover, mastoid air cells also open into the middle ear cavity, and their mucous lining is also continuous with the mucous lining of the middle ear. So when a middle ear infection develops, it may spread to produce *mastoiditis*.

The *internal* ear has two parts. One is made of bone, the other of a membrane that lies inside the bone. Both have complicated shapes, and for this reason are called *labyrinths*. Each labyrinth has three parts: *vestibule, semicircular canals,* and *cochlea* (Figure 4-10). The semicircular canals are three half circles; the cochlea is shaped like a snail shell, which is what the word cochlea means.

Two special sense organs for two different kinds of sensations—hearing and balance—are located in the internal ear. The hearing sense organ, which lies inside the cochlea, is called the *organ of Corti*. There are two balance or equilibrium sense organs—one called the *macula*, which is located in the vestibule of the inner ear, and the other called the *crista ampullaris*. The crista is located in the semicircular canals.

Outline summary—The nervous system

CELLS OF NERVOUS SYSTEM
Two kinds of cells found only in organs of nervous system—neurons and neuroglia
1. Neuroglia—connective tissue cells of two main types
   a. Astrocytes—star-shaped cells which anchor small blood vessels to neurons
   b. Microglia—small cells which move about in inflamed brain tissue carrying on phagocytosis, that is, they engulf and destroy microorganisms and other injurious particles
2. Neurons or nerve cells
   a. Consist of three main parts: dendrites —conduct impulses to cell body of neuron; cell body of neuron, and axon— conducts impulses away from cell body of neuron
   b. Neurons classified according to function as sensory—conduct impulses to spinal cord and brain; motoneurons— conduct impulses away from brain and cord out to muscles and glands; and interneurons—conduct impulses from sensory neurons to motoneurons

Conduction of nerve impulses
1. Nerve impulses are conducted from receptors to effectors over neuron pathways or reflex arcs; conduction by reflex arc results in a reflex, that is, contraction by a muscle or secretion by a gland
2. Simplest reflex arcs called 2-neuron arcs— consist of sensory neurons synapsing in spinal cord with motoneurons; 3-neuron arcs consist of sensory neurons synapsing in spinal cord with interneurons, which synapse with motoneurons

BRAIN AND CORD COVERINGS AND FLUID SPACES
1. Coverings—cranial bones, vertebrae, and meninges
2. Fluid spaces—ventricles inside brain, central canal inside cord, and spaces in arachnoid membrane (middle layer of meninges) around both brain and cord

ORGANS OF NERVOUS SYSTEM
1. Central nervous system (CNS)—brain and spinal cord
2. Peripheral nervous system (PNS)—all nerves

Brain
Cerebrum
1. Largest part of human brain
2. Outer layer called cerebral cortex; made up of lobes, which are made up of convolutions; cortex composed mainly of cell bodies of neurons
3. Interior of cerebrum composed mainly of nerve fibers arranged in bundles called tracts
4. Functions of cerebrum—mental processes of all types, including sensations, con-

sciousness, and voluntary control of movements

Cerebellum

1. Second largest part of human brain
2. Helps control muscle contractions so that they produce coordinated movements so that we can maintain balance, move smoothly, and sustain normal postures

Medulla

1. Lowest part of brain, an enlarged extension of spinal cord in cranial cavity
2. Most vital part of brain in that it controls heartbeat, blood pressure, and respirations

Thalamus

Sensory impulses pass through thalamus on way to cerebral cortex; thalamus probably responsible for emotions associated with sensations *FUNCTIONS AS A RELAY STATION FOR SENSORY IMPULSES*

Hypothalamus

1. Acts as the major center for controlling autonomic nervous system; therefore helps control the functioning of most internal organs
2. Controls hormone secretion by both anterior and posterior pituitary glands; therefore indirectly helps control hormone secretion by most other endocrine glands
3. Acts as center for controlling appetite; therefore helps regulate amount of food eaten and body weight
4. Functions in some way to maintain the waking state
5. Probably contains reward and punishment centers

Spinal cord

1. Structure—outer part composed of many tracts; interior composed mainly of cell bodies of neurons
2. Functions
   a. Conducts sensory impulses to brain and motor impulses from brain
   b. Serves as center for many reflexes

Cranial nerves

See Table 4-1, page 57

Spinal nerves

1. Structure—contain dendrites of sensory neurons and axons of motor neurons
2. Functions—conduct impulses
   a. Necessary for sensations
   b. Necessary for voluntary movements

AUTONOMIC (INVOLUNTARY) NERVOUS SYSTEM

1. Autonomic neurons are motoneurons that conduct impulses to visceral effectors (smooth muscle, cardiac muscle, and glands)
2. Divisions—sympathetic and parasympathetic; both sympathetic and parasympathetic neurons conduct impulses to heart and most smooth muscle and glands
3. Functions
   a. In general, sympathetic and parasympathetic impulses tend to produce opposite effects (see Table 4-2, page 58); together, they help control heart beat, contraction of smooth muscle, and secretion of glands
   b. Sympathetic impulses dominate control of involuntary, automatic functions in times of stress, causing them to respond in ways that make possible maximum muscular activity; sympathetic impulses prepare body for "fight or flight"
   c. Parasympathetic impulses dominate control of most involuntary functions under normal conditions

SENSE ORGANS

1. All receptors (beginning of dendrites of sensory neurons) are sense organs
2. Special sense organs—eye and ear
3. Kinds of senses—many more than the familiar five senses; for example, several kinds of touch senses; proprioception—sense of position and movement

The eye

1. Coats of eyeball
   a. Sclera—tough outer coat; whites of eye; cornea is transparent part of sclera over iris
   b. Chorioid—front part of this coat made up of ciliary muscle and iris, the colored part of eye; pupil is hole in center of iris; contraction of iris muscle dilates or constricts pupil
   c. Retina—innermost coat of eye; contains rods (receptors for night vision) and cones (receptors for day vision and color vision)
2. Conjunctiva—mucous membrane that covers front surface of eyeball and lines lids
3. Lens—transparent body behind pupil; focuses light rays on retina
4. Eye fluids
   a. Aqueous humor—in anterior cavity in front of lens
   b. Vitreous humor—in posterior cavity behind lens

The ear
1. Parts
   a. External ear—consists of pinna and auditory canal
   b. Middle ear—contains auditory bones (malleus, incus, and stapes); lined with mucous membrane; eustachian tubes and mastoid sinuses open into middle ear
   c. Internal ear or labyrinth—vestibule, semicircular canals, and cochlea are three divisions of internal ear; organ of Corti, sense organ of hearing, lies in cochlea; contains sense organs of equilibrium (macula and crista); macula lies in vestibule and crista lies in semicircular canals

Review questions—The nervous system

1. What general function does the nervous system perform?
2. What other system performs the same general function as the nervous system?
3. What is each of the following?
   dendrite   tract
   neuron   synapse
   nerve
4. Define sensory neuron, motor neuron, and central neuron.
5. Explain briefly the meaning of reflex arc and reflex.
6. What does "CNS" mean?
7. What are the meninges?
8. What general functions does the cerebrum perform?
9. What general functions does the cerebellum perform?
10. What general functions does the spinal cord perform?
11. What general functions do spinal nerves perform?
12. What are some of the functions performed by cranial nerves?
13. Why is the medulla considered the most vital part of the brain?
14. What general functions does the autonomic nervous system perform?
15. How do the sympathetic and parasympathetic nervous systems differ in function?
16. Define receptors and effectors.
17. What does proprioception or kinesthesia mean?
18. Describe as fully as you can the structure of the eye.
19. Describe as fully as you can the structure of the ear.
20. What is each of the following?
   cornea   iris
   organ of Corti   retina

How the body distributes food and oxygen and removes wastes

V

The circulatory system

Have you ever thought what would happen if all the transportation ceased in your city or town? How soon you would have no food to eat and how soon rubbish and waste would pile up, for instance? Stretch your imagination just a little and you can visualize many disastrous results. Lack of food transportation alone would threaten every individual's life. Similarly, cells, the individuals of the body, face a threat to survival if transporation within the body ceases. The system which provides this vital transportation service is the circulatory system. Its organs are the heart, blood vessels, lymphatic vessels, and lymph nodes. The fluids circulated are blood and lymph. We shall begin this chapter with a discussion of blood.

BLOOD STRUCTURE AND FUNCTIONS

Blood is one of the three main fluids of the body. (The fluid around cells and the fluid inside them are the other two.) The liquid part of blood is called *plasma*. Floating in the plasma are hundred of millions of blood cells.

There are three main types and several subtypes of blood cells as follows:
1. Red cells or erythrocytes
2. White cells or leukocytes
    a. Granular leukocytes (have granules in their cytoplasm)
       (1) Neutrophils
       (2) Eosinophils
       (3) Basophils
    b. Nongranular leukocytes (do not have granules in their cytoplasm)
       (1) Lymphocytes
       (2) Monocytes

Examine Figure 5-1 to see what each of these different kinds of blood cells looks like under the microscope.

The number of blood cells in the body is hard to believe. For instance, 5 million

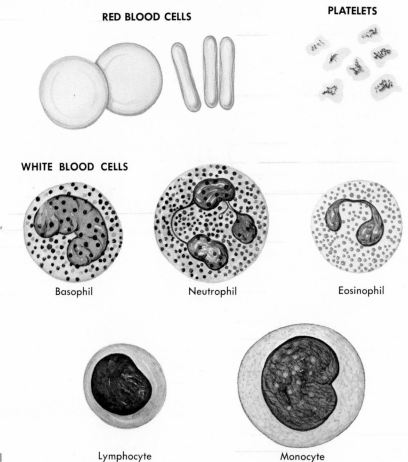

RED BLOOD CELLS

PLATELETS

WHITE BLOOD CELLS

Basophil   Neutrophil   Eosinophil

Fig. 5-1   Lymphocyte   Monocyte

Human blood cells. There are close to 30 trillion blood cells in the adult. Each cubic millimeter of blood contains from 4½ to 5½ million red blood cells and an average total of 7,500 white blood cells.

red cells and 7,500 white blood cells in 1 cubic millimeter of blood (a drop only about 1/25th of an inch long and wide and high) would be considered a normal "red count" and a normal "white count." Since both red and white cells are continually being destroyed, the body has to continually make new ones to take their place at a really staggering rate—a few million red cells alone each second.

Two kinds of connective tissue—*myeloid tissue* and *lymphatic tissue*—make blood cells for the body. A better known name for myeloid tissue is red bone marrow. In the adult body it is present chiefly in the sternum, ribs, and cranial bones although a few other bones also contain small amounts of this valuable substance. Red bone marrow forms all types of blood cells except lymphocytes and monocytes. These are formed by lymphatic tissue located chiefly in the lymph nodes and the spleen.

Sometimes for one reason or another, the blood-forming tissues cannot maintain normal numbers of blood cells. The number of red cells, for example, frequently drops below normal, a condition called *anemia. Leukopenia,* or an abnormally low white count, occurs only occasionally. An abnormally high white count, *leukocytosis*, on the other hand, is quite common, since it almost always accompanies infections. There is also a malignant disease, *leukemia*, in which the number of white cells increases tremendously; you may have heard of this disease as "blood cancer."

Blood cells perform several important functions. Red cells transport oxygen to all the other cells in the body. A chemical in them called *hemoglobin* unites with oxygen to form oxyhemoglobin. If hemoglobin falls below the normal level—as it does in anemia—it starts an unhealthy chain reaction: less hemoglobin—less oxygen transported to cells—slower catabolism by cells—less energy supplied to cells—decreased cellular functions. If you understand this relation between hemoglobin and energy, you can guess correctly what an anemic person's chief complaint will probably be—that he feels "so tired all the time." Red blood cells perform one other function; besides transporting oxygen, they also transport carbon dioxide.

White blood cells carry on a function perhaps slightly less vital than red cells, but one, nevertheless, that often has life-saving importance. They defend the body from microorganisms that have succeeded in invading the tissues or blood stream. Neutrophils and monocytes, for example, engulf microbes. They actually take them into their own cell bodies and digest them. This process is called *phagocytosis*, and the cells that carry it on are called *phagocytes*. Most numerous of the phagocytes are the neutrophils, but in addition to neutrophils and monocytes, the body also has other phagocytes that are not white blood cells. They are classed as reticulo-endothelial cells—a type of connective tissue cells. Microglia (page 48), for example, are reticulo-endothelial phagocytes.

Lymphocytes, it is now thought, also help protect us against infections. But they do it by a process different from phago-

cytosis. Lymphocytes are believed to function in the immune mechanism, the process which makes us immune to infectious diseases. According to present ideas, the immune mechanism starts to operate when microbes invade the body. In some way or another their presence acts to stimulate lymphocytes to start multiplying and to become transformed into plasma cells. Presumably, each kind of microbe can stimulate only one specific kind of lymphocyte to multiply and form one kind of plasma cell. The kind of plasma cell formed is the kind that can make a specific kind of antibody. The kind of antibody made is the one that can destroy the particular kind of microbe that has invaded the body in the first place and set the immune mechanism in operation.

Blood plasma

Blood plasma is the liquid part of the blood, or blood minus its cells. It consists of water with many substances dissolved in it. All of the chemicals needed by cells to stay alive have to be brought to them by the blood. Food, therefore, and salts are dissolved in plasma. So, too, is a small amount of oxygen. (Most of the oxygen in the blood is carried in the red blood cells as oxyhemoglobin.) Wastes that cells need to get rid of are dissolved in plasma and transported to the excretory organs. And, finally, plasma contains dissolved in it the hormones that help control our cells' activities and the antibodies that help protect us against microorganisms.

Many people seem curious about how much blood they have. The amount depends upon how big they are and whether they are male or female. A big person has more blood than a small person and a man has more blood than a woman. But as a general rule, most adults probably have between 4 and 6 quarts (4,000 and 6,000 milliliters) of blood.

The volume of the plasma part of blood is usually a little more than half the volume of whole blood. Blood cells make up the remaining part of whole blood's volume. Examples of normal volumes are the following: plasma volume—2,600 milliliters; blood cell volume—2,400 milliliters; total blood volume—5,000 milliliters.

If you read many advertisements or watch many television commercials, you may think that almost everyone has "acid blood" at some time or other. Nothing could be further from the truth. Blood is alkaline; it rarely reaches even the neutral point. Not only that—if the alkalinity of your blood decreases toward neutral, you are a very sick person; in fact, you have what is called *acidosis*. But even in this condition, blood almost never becomes the least bit acid; it just becomes less alkaline than normal.

Blood types (or blood groups)

Before we discuss blood types, we need to define the terms "antigens" and "antibodies." An *antigen* is a substance that can stimulate the body to make antibodies. Almost all substances that act as antigens are foreign proteins. In other words, they are not the body's own natural proteins, but are proteins that have entered the body from the outside—by injection or transfusion or some other method.

The word "antibody" can be defined in terms of what causes its formation or in terms of how it functions. Defined the first way, an *antibody* is a substance made by the body in response to stimulation by an antigen. Defined according to its functions, an antibody is a substance that reacts in some way with the antigen that stimulated its formation. Many antibodies, for example, clump (or agglutinate) their antigens; that is, they cause them to stick together in little clusters.

Every person's blood belongs to one of the following four blood types: type A, type B, type AB, or type O. Suppose that you have type A blood (as do about 41% of Americans). What does this mean? Well,

the letter A stands for a certain type of antigen (a protein) present in your red blood cells when you were born. Because you were born with type A antigen, your body does not form antibodies to react with it. In other words, your blood plasma contains no anti-A antibodies. It does contain anti-B antibodies, however. For some unknown reason, these antibodies are present naturally in type A blood. The body did not form them in response to the presence of their antigen. In summary, then, in type A blood, the red cells contain type A antigen and the plasma contains anti-B antibodies.

In type AB blood, as its name indicates, the red cells contain both type A and type B antigens and the plasma contains neither anti-A nor anti-B antibodies. The opposite is true of type O blood—its red cells contain neither type A nor type B antigens and its plasma contains both anti-A and anti-B antibodies.

Bad effects or even death can result from a blood transfusion if the donor's red cells become agglutinated by antibodies in the recipient's plasma. If a donor's red cells do not contain any A or B antigen, they of course cannot be clumped by anti-A or anti-B antibodies. For this reason the type blood which contains neither A nor B antigens—namely, type O blood—can be used as donor blood without the danger of anti-A or anti-B antibodies clumping its red cells. Type O blood is therefore called "*universal donor*" *blood*. "*Universal recipient*" *blood* is type AB; it contains no anti-A or anti-B antibodies in its plasma, and so cannot clump any donor's red cells containing A or B antigens.

In recent years, the expression "*Rh positive*" blood has become a familiar one. It means that the red blood cells of this type blood contain an antigen called the Rh factor. If, for example, a person has type AB, Rh positive blood, his red cells contain type A antigen, type B antigen, and the Rh factor.

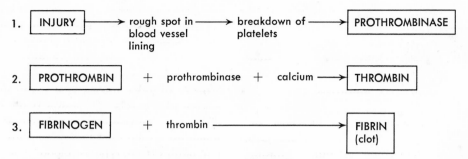

Fig. 5-2

Diagram showing the main steps in blood clotting—a process far more complex than is indicated here.

In *Rh negative* blood, the red cells do not contain the Rh factor nor does the plasma contain anti-Rh antibodies naturally. But if Rh positive blood cells are introduced into an Rh negative person's body, anti-Rh antibodies soon appear in his blood plasma. In this fact lies the danger for a baby born to an Rh negative mother and Rh positive father. If the baby is Rh positive, the mother's body may be stimulated to form anti-Rh antibodies. If then she later carries another Rh positive fetus, it may develop a disease called erythroblastosis fetalis, due to the mother's Rh-antibodies reacting with the baby's Rh positive cells.

BLOOD CLOTTING

Your life might someday be saved just because your blood can clot. A clot plugs up torn or cut vessels and so stops bleeding that otherwise might prove fatal. The story of how blood clots is the story of a rapid-fire reaction. The first step in the chain is some kind of an injury to a blood vessel which makes a rough spot in its lining. (Normally the lining of blood vessels is extremely smooth.) Almost immediately some of the blood platelets break up as they flow over the rough spot in the vessel's lining and release a substance into the blood which leads to the formation of another substance called *prothrombinase, thromboplastin,* or *thrombokinase.* (Figure 5-2.) In the next step prothrombinase combines with prothrombin (a protein present in normal blood), calcium, and other sub-

stances to form *thrombin.* Then in the last step thrombin reacts with fibrinogen (another protein present in normal blood) to change it to a gel called fibrin. Under the microscope fibrin looks like a tangle of fine threads with red blood cells caught in the tangle. The red cells give the red color to clotted blood.

The clotting mechanism as just described contains clues for ways of stopping bleeding by speeding up blood clotting. For example, you might simply apply gauze to a bleeding surface. Its slight roughness would cause more platelets to break down and release more prothrombinase. The additional prothrombinase would then make the blood clot more quickly.

Physicians frequently prescribe vitamin K before surgery to make sure that the patient's blood will clot fast enough to prevent hemorrhage. Figure 5-3 shows the somewhat roundabout way in which vitamin K acts to hasten blood clotting.

Unfortunately clots sometimes form in uncut blood vessels of the heart, brain, lungs, or some other organ—a dreaded thing, because they may produce sudden death by shutting off the blood supply to a vital organ. When a clot stays in the place where it has formed, it is called a *thrombus* and the condition is spoken of as *thrombosis.* If part of the clot dislodges and circulates through the bloodstream, the dis-

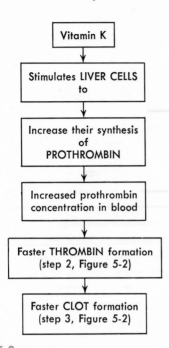

Fig. 5-3

Diagram showing how vitamin K acts to accelerate blood clotting.

lodged part is then called an *embolus* and the condition is called an *embolism.* Suppose that your doctor told you that you had a clot in one of your coronary arteries. Which diagnosis would he make—coronary thrombosis or coronary embolism—if he thought that the clot had formed originally in one of these small vessels that supply blood to heart muscle cells? Doctors now have some drugs that they can use to help prevent thrombosis and embolism. Dicumarol is one. It blocks the stimulating effect of vitamin K on the liver and consequently the liver cells make less prothrombin. The blood prothrombin content soon falls low enough to prevent clotting. (See Figure 5-3.)

THE HEART

No one needs to be told where his heart is nor what it does. Everyone knows that the heart is in the chest, that it beats night and day to keep the blood flowing, and that if it stops life stops. Most of us probably think of the heart as located on the left side, but as you can see in Figure 1-2, it occupies the lower portion of the mediastinum with most of its mass to the left of the midline of the body. The *apex* (blunt point of the lower edge of the heart) lies on the diaphragm, pointing toward the left. To count the apical beat, one must place a stethoscope directly over the apex, that is, in the space between the fifth and sixth ribs on a line with the midpoint of the left clavicle.

If you cut open a heart you can see many of its main structural features. (See Figure 5-4. It is hollow, not solid. A partition (the septum) divides it into right and left sides. It has four cavities inside: a small upper cavity *(atrium)* and a larger lower cavity *(ventricle)* on each side. Cardiac muscle tissue composes the wall of the heart; it is usually referred to as the *myocardium.* (Myocarditis, therefore, means inflammation of the heart muscle.) A very smooth tissue, *endocardium,* lines the cavity of the heart. Endocarditis, of course, means inflammation of the heart lining. This condition can cause rough spots to develop in the endocardium which may lead to thrombosis. (Why? Look back at Figure 5-2 if you are not sure.)

The heart has a covering as well as a lining. Its covering, the *pericardium,* consists of two layers of fibrous tissue with a small space in between. The inner layer of the pericardium covers the heart like an apple skin covers an apple, but the outer layer fits around the heart like a loose-fitting sack, allowing enough room for the heart to beat in it. These two pericardial layers slip against each other without friction when the heart beats because they are moist, not dry surfaces. (Have you ever walked carefully on a wet floor? If so, you were consciously or unconsciously using your knowledge of the principle that moist surfaces are slippery.) A thin film of peri-

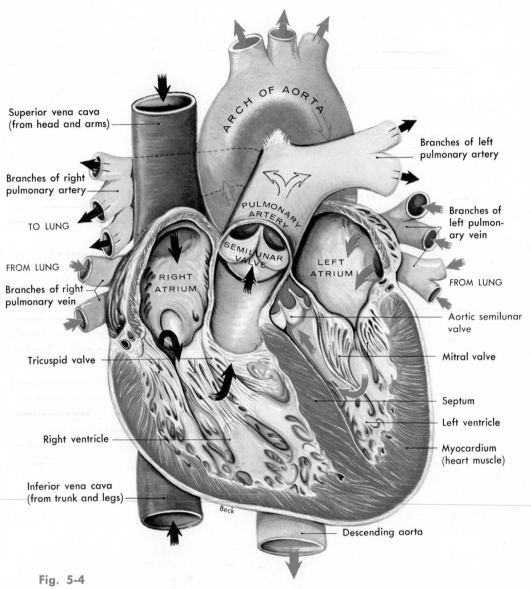

Superior vena cava
(from head and arms)

ARCH OF AORTA

Branches of left
pulmonary artery

Branches of right
pulmonary artery

PULMONARY
ARTERY

Branches of
left pulmon-
ary vein

TO LUNG

SEMILUNAR
VALVE

FROM LUNG

RIGHT
ATRIUM

LEFT
ATRIUM

FROM LUNG

Branches of right
pulmonary vein

Aortic semilunar
valve

Tricuspid valve

Mitral valve

Septum

Left ventricle

Right ventricle

Myocardium
(heart muscle)

Inferior vena cava
(from trunk and legs)

Beck

Descending aorta

Fig. 5-4

Cutaway view of the front section of the heart showing the four chambers, the valves, the
openings, and the major vessels. Arrows indicate the direction of blood flow; the black arrows
represent unoxygenated blood and the red arrows oxygenated blood. The two branches of the
right pulmonary vein extend from the right lung behind the heart to enter the left atrium.

cardial fluid furnishes the lubricating moist-
ness between the heart and its enveloping
pericardial sac.

Four valves keep blood flowing through
the heart in the right direction. The *tricuspid valve,* at the opening between the

right atrium and the right ventricle, lets
blood flow from the atrium into the ventricle but prevents it from flowing in the
opposite direction. The *mitral* or *bicuspid
valve,* at the opening between the left
atrium and the left ventricle, serves the

ARTERY

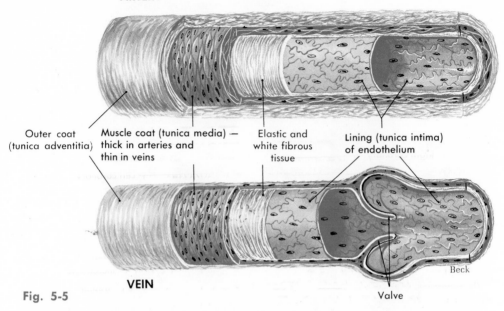

Outer coat (tunica adventitia)

Muscle coat (tunica media) — thick in arteries and thin in veins

Elastic and white fibrous tissue

Lining (tunica intima) of endothelium

Beck

VEIN

Fig. 5-5   Valve

Schematic drawings of an artery and vein showing comparative thicknesses of the three coats: the outer coat (tunica adventitia), the muscle coat (tunica media), and the lining of endothelium (tunica intima). Note that the muscle and outer coats are much thinner in veins than in arteries and that veins have valves.

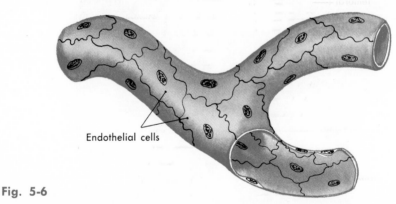

Endothelial cells

Fig. 5-6

The walls of capillaries consist of only a single layer of endothelial cells. These thin, flattened cells permit the rapid movement of substances between blood and interstitial fluid. Note that capillaries have no smooth muscle layer, elastic fibers, or surrounding adventitia.

same purpose for the left side of the heart. The *pulmonary semilunar valve* at the beginning of the pulmonary artery allows blood to flow out of the right ventricle but prevents it from backflowing into the ventricle. The *aortic semilunar valve* at the beginning of the aorta allows blood to flow out of the left ventricle up into the aorta but prevents backflow into this ventricle.

THE BLOOD VESSELS

Arteries, veins, and capillaries—these are the names of the three main kinds of blood

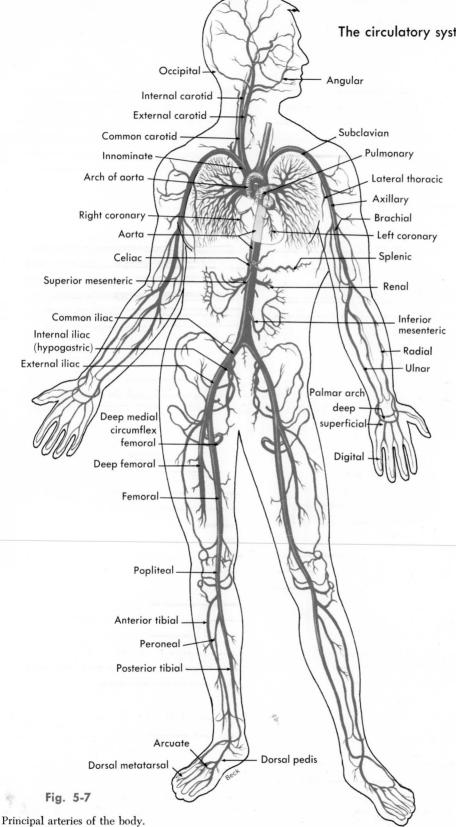

Occipital
Angular
Internal carotid
External carotid
Common carotid
Subclavian
Innominate
Pulmonary
Arch of aorta
Lateral thoracic
Axillary
Right coronary
Brachial
Aorta
Left coronary
Celiac
Splenic
Superior mesenteric
Renal
Common iliac
Inferior mesenteric
Internal iliac (hypogastric)
Radial
External iliac
Ulnar
Palmar arch
deep
superficial
Deep medial circumflex femoral
Deep femoral
Digital
Femoral
Popliteal
Anterior tibial
Peroneal
Posterior tibial
Arcuate
Dorsal metatarsal
Dorsal pedis
Beck

Fig. 5-7

Principal arteries of the body.

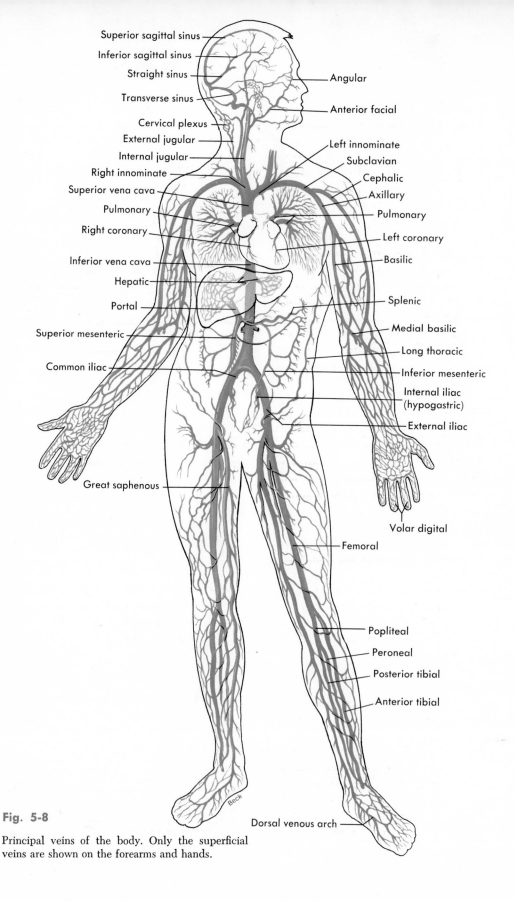

Superior sagittal sinus
Inferior sagittal sinus
Straight sinus
Transverse sinus
Cervical plexus
External jugular
Internal jugular
Right innominate
Superior vena cava
Pulmonary
Right coronary
Inferior vena cava
Hepatic
Portal
Superior mesenteric
Common iliac
Great saphenous

Angular
Anterior facial
Left innominate
Subclavian
Cephalic
Axillary
Pulmonary
Left coronary
Basilic
Splenic
Medial basilic
Long thoracic
Inferior mesenteric
Internal iliac (hypogastric)
External iliac
Volar digital
Femoral
Popliteal
Peroneal
Posterior tibial
Anterior tibial
Dorsal venous arch

Beck

Fig. 5-8

Principal veins of the body. Only the superficial
veins are shown on the forearms and hands.

vessels. *Arteries* carry blood away from the heart toward capillaries. *Veins* carry blood toward the heart away from capillaries. *Capillaries* carry blood from tiny arteries (*arterioles*) into tiny veins (*venules*). When a surgeon cuts into the body, he can see arteries, arterioles, veins, and venules. Capillaries he cannot see because they are microscopic-sized vessels. An important point about the structure of a capillary is that its wall (that is, the capillary membrane) is very thin—only one cell thick to be exact. Because a single layer of flat epithelial-like cells composes the capillary membrane, foods, oxygen, wastes, and other substances can quickly pass through it on their way to or from cells. Figures 5-5 and 5-6 illustrate the structure of arteries, capillaries, and veins.

The largest artery in the body is the *aorta*. The largest vein is the *vena cava*. The aorta carries blood out of the left ventricle of the heart, and the vena cava returns blood to the right atrium after the blood has circulated through the body. Study Figure 5-7 to find out the names of the main arteries of the body and Figure 5-8 for the names of the main veins. Then see whether you can answer these questions: what is the name of the main artery of the thigh? of the upper arm? of the thumb side of the lower arm? Neck arteries are named what? neck veins? Answer question 23, page 87. Make up some more questions to quiz yourself about blood vessel names and locations.

CIRCULATION

The word circulation implies that something moves, that it moves over a circular route, and that it moves over this route repeatedly. Applied to blood, circulation means the movement of blood over and over again through vessels that form a circular route. (By circular route, we mean one that begins and ends at the same place, not necessarily one shaped like a circle.) Since the heart pumps the blood and since

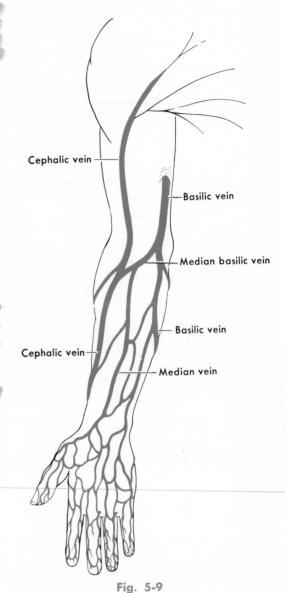

Cephalic vein

Basilic vein

Median basilic vein

Basilic vein

Cephalic vein

Median vein

Fig. 5-9

Main superficial veins of the arm.

the blood returns to the right atrium of the heart when it has completed one circuit through the blood vessels, we shall consider that circulation starts in the right atrium. Examine Figure 5-10 carefully. Imagine that you have injected a drug into a patient's right upper arm. Soon the drug would diffuse into the blood in the capil-

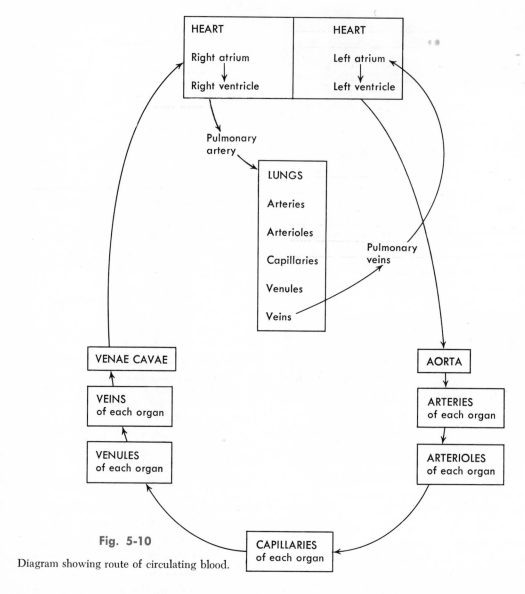

Fig. 5-10

Diagram showing route of circulating blood.

laries of the upper arm. Look again at Figure 5-10 to find out what kind of vessels the blood would flow into as it left the arm's capillaries. Notice next what two organs the drug-containing blood would have to pass through before it could enter the capillaries of any other organ in the body. First it would have to return to the heart—which side? which chamber?—next it would have to circulate through the lungs, come back to the heart (which side? which chamber?), and then be pumped by the left ventricle out into the aorta and on into arteries, arterioles, and capillaries of other organs. Make sure that you understand the blood's circulation route by answering questions 14 to 18 on page 86 and checking your answers with Figure 5-10.

As blood flows through capillaries in the lungs, it changes from venous blood to arterial blood by unloading carbon dioxide and picking up oxygen. Its color changes in the process from the deep crimson that identifies venous blood to the bright scarlet of arterial blood. Then as blood flows through tissue capillaries (all capillaries except those in the lungs), it changes back from arterial to venous blood by oxygen leaving the blood to enter cells and carbon dioxide leaving the cells to enter blood.

Blood flows into the heart muscle itself by way of two small vessels which are surely the most famous of all the blood vessels—the coronary arteries—famous because coronary disease kills so many thousands every year. The coronary arteries are the aorta's first branches. The openings into these small vessels lie behind the flaps of the aortic semilunar valves. In both coronary thrombosis and coronary embolism (pages 71 to 72), a blood clot occludes or plugs up some part of a coronary artery. Blood cannot pass through the occluded vessel and so cannot reach the heart muscle cells it normally supplies. Deprived of oxygen, these cells soon die; or to use the medical term, myocardial infarction occurs. Myocardial infarction is a common cause of death in middle-aged and old people.

The term *"portal circulation"* refers to the following route of blood flow through the liver. Veins from the abdominal digestive organs (stomach, pancreas, and intestines) and from the spleen empty into the portal vein; its branches empty into arterioles which empty into the capillaries of the liver. Blood leaves the liver by way of the hepatic vein which drains into the inferior vena cava. How does the portal system differ from circulation through other organs? Does venous blood ordinarily flow into capillaries? See Figure 5-10 if you are not sure about this. The main reason for the detour of venous blood through the liver before its return to the heart is probably the process of liver glycogenesis (page 93).

Circulation in a baby before birth is not the same as after birth, mainly because before birth the baby's blood has to get oxygen and food from the mother's blood in-

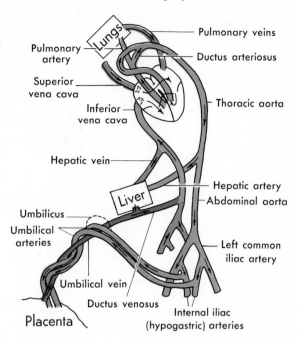

Fig. 5-11

Diagram of fetal circulation. Note these six structures, which are present in a baby's body at birth, but not in a normal adult body: (1) umbilical arteries (two of them); (2) placenta (afterbirth); (3) umbilical vein; (4) ductus venosus; (5) ductus arteriosus; (6) foramen ovale (the opening—not labeled—between the right and left atria of the heart).

stead of from its own lungs and digestive tract. Therefore relatively little blood circulates through the baby's lungs and digestive organs before it is born, but all of the baby's blood circulates through the placenta (afterbirth) where the exchange of substances with the mother's blood occurs. Figure 5-11 shows the plan of fetal circulation.

BLOOD PRESSURE

Perhaps a good way for us to try to understand blood pressure is to try to answer the questions what? where? why? and how? about it. What is blood pressure? Just what the words say—blood pressure is the pressure or push of blood.

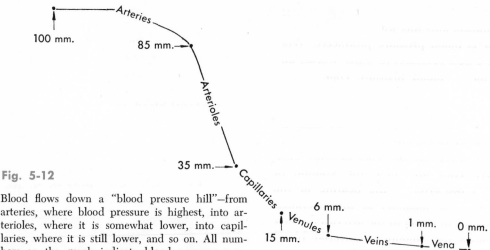

Fig. 5-12

Blood flows down a "blood pressure hill"—from arteries, where blood pressure is highest, into arterioles, where it is somewhat lower, into capillaries, where it is still lower, and so on. All numbers on the graph indicate blood pressure measured in millimeters of mercury. The top figure, 100 mm., represents the average pressure in the aorta.

Where? Blood pressure exists in all blood vessels, but it is highest in the arteries and lowest in the veins. In fact if we list blood vessels in order according to the amount of blood pressure in them and draw a graph of this, as in Figure 5-12, the graph looks like a hill, with aortic blood pressure at the top and vena caval pressure at the bottom. This blood pressure hill is spoken of as the blood pressure gradient, or to be more exact, the term *blood pressure gradient* means the difference between two blood pressures. The blood pressure gradient for the entire systemic circulation is the difference between the average or mean blood pressure in the aorta and the blood pressure at the termination of the venae cavae where they join the right atrium of the heart. If you use the typical normal figures shown in Figure 5-12, the systemic blood pressure gradient figures out to be 100 millimeters of mercury pressure.

More important for you to remember than the amount of the blood pressure gradient is its function. The blood pressure gradient keeps blood flowing. When a blood pressure gradient is present, blood circulates. And conversely, when a blood pressure gradient is not present, blood does not circulate. For example, suppose that the blood pressure in the arteries were to decrease so that it became equal to the average pressure in arterioles. There would no longer be a blood pressure gradient between arteries and arterioles, and therefore there would no longer be a force to move blood out of arteries into arterioles. Circulation would stop, in other words, and very soon life itself would cease. This is why when arterial blood pressure is observed to be falling rapidly, whether in surgery or elsewhere in a hospital, emergency measures are quickly started to try to reverse this fatal trend.

What we have just said in the preceding paragraph may start you wondering why high blood pressure (meaning, of course, high arterial blood pressure) is not good for circulation and low pressure bad for it. High blood pressure is considered "bad" for several reasons. For one thing, if it becomes too high, it may cause the rupture of some of the delicate-walled capillaries (for example, in the brain, as happens in a stroke). But low blood pressure also, as we have already seen, can be dangerous. If arterial pressure falls low enough, circulation and life cease. Massive hemorrhage, for instance, kills this way.

How is blood pressure produced? What causes blood pressure, in other words, and what makes blood pressure change from time to time? To give a brief and at the same time a true answer to these questions seems almost impossible, but we shall try to do so by discussing only the three factors that affect arterial blood pressure most directly. They are these—the volume of blood in the arteries, the beating of the heart, and the diameter of the arterioles. The volume of blood in the arteries directly determines the amount of arterial blood pressure. In general, the more blood in the arteries, the higher the blood pressure. Conversely, the less blood in the arteries the lower the blood pressure tends to be. Hemorrhage demonstrates well this relationship between blood volume and blood pressure. In hemorrhage, a marked loss of blood occurs and this decrease in the volume of blood causes blood pressure to drop. In fact the major sign of hemorrhage is a decrease in the blood pressure.

Both the strength and the rate of the heartbeat affect blood pressure. Each time the left ventricle contracts, it squeezes a certain volume of blood into the aorta and on into other arteries. The stronger each contraction is, of course the more blood it pumps into the aorta and arteries. Conversely, the weaker each contraction is, the less blood it pumps. Suppose that one contraction of the left ventricle pumps 70 milliliters (about one-third of a cupful) of blood into the aorta and suppose that the heart beats 70 times a minute. Seventy milliliters times 70 equals 4,900 milliliters. Almost five quarts of blood, in other words, would enter the aorta and arteries every minute. Now suppose that the heartbeat were to become weaker. Suppose that each contraction of the left ventricle pumps only 50 milliliters of blood instead of 70 into the aorta. If the heart still contracts only 70 times a minute, it will obviously pump

much less blood into the aorta—only 3,500 milliliters instead of the more normal 4,900 milliliters per minute. This decrease in the heart's output of blood tends to decrease the volume of blood in the arteries, and the decreased arterial blood volume tends to decrease arterial blood pressure. In summary, the strength of the heartbeat affects blood pressure this way—a stronger heartbeat tends to increase blood pressure and a weaker beat tends to decrease it.

Now let us turn our attention to the rate of the heartbeat and its effect on arterial blood pressure. We might expect that when the heart beats faster, more blood would enter the aorta and that therefore the arterial blood volume and blood pressure would increase. But this does not necessarily happen. Why? Because often when the heart beats faster, each contraction of the left ventricle takes place so rapidly that it squeezes out less blood than usual into the aorta. If, for example, the heart rate speeds up to 100 times a minute, each beat might pump only 40 milliliters of blood into the aorta instead of the normal 70 milliliters. What do you think would happen to blood pressure then? Knowing what you now do about the relation between the volume of blood pumped per minute by the heart, the volume of blood in the arteries, and the blood pressure, do you think that the blood pressure would increase, or decrease, or stay the same? Both arterial blood volume and blood pressure would tend to decrease under these conditions even though the heart rate had increased. What generalization, then, can we make? We can only say that an increase in the rate of the heartbeat *tends* to increase blood pressure and a decrease in the rate tends to decrease blood pressure. But whether a change in the heart rate actually produces a similar change in blood pressure, depends on whether the strength of the heart's beat also changes, and in which direction.

Another factor that we ought to mention

in connection with blood pressure is the viscosity of blood, or in plainer language, its stickiness. If blood becomes less viscous than normal, blood pressure decreases. For example, if a person suffers a hemorrhage, fluid will move into his blood from his interstitial fluid. This dilutes the blood and decreases its viscosity, and blood pressure then falls because of the decreased viscosity. As you may know, after hemorrhage either whole blood or plasma is preferred to saline solution for transfusions. The reason is that saline solution is not a viscous liquid, and so cannot keep blood pressure at a normal level.

Blood pressure does not stay the same every minute. It changes from time to time even when we are healthy. For example, when we exercise strenuously, our blood pressure goes up. Not only is this normal but the increased blood pressure serves a good purpose. It tends to increase circulation, to bring more blood to muscles each minute, and thus to supply them with more oxygen and food for more energy.

A normal average arterial blood pressure is 120/80, or 120 millimeters of mercury systolic pressure (as the ventricles contract) and 80 millimeters of mercury diastolic pressure (as the ventricles relax).

PULSE

What you feel when you take a pulse is an artery expanding and then recoiling alternately. In order to feel a pulse you must place your fingertips over an artery that lies near the surface of the body and over a bone or other firm background. Six such places are given below. (You can locate all of them on Figure 5-7, page 75, except the temporal and facial arteries.)

1. *Radial artery*—at the wrist
2. *Temporal artery*—in front of the ear or above and to the outer side of the eye
3. *Common carotid artery*—in the neck along the front edge of the sternocleidomastoid muscle at the level of

the lower margin of the thyroid cartilage
4. *Facial artery*—at the lower margin of the lower jaw bone on a line with the corners of the mouth
5. *Brachial artery*—at the bend of the elbow along the inner margin of the biceps muscle.
6. *Dorsalis pedis artery*—on the front surface of the foot, just below the bend of the ankle joint.

THE LYMPHATIC SYSTEM

The lymphatic system is not really a separate system of the body. It is part of the circulatory system since it consists of a moving fluid—lymph—which comes from the blood and returns to the blood by way of the lymphatic vessels. Lymph forms in this way: blood plasma filters out of the capillaries into the microscopic spaces between tissue cells. Here the liquid is called *interstitial fluid* or tissue fluid. Some of the interstitial fluid goes back into the blood by the same route it came out, that is, back through the capillary membrane. But most of the interstitial fluid enters tiny lymphatic capillaries to become lymph. It next moves on into larger lymphatics, and finally enters the blood in veins in the neck region. The largest lymphatic vessel is the *thoracic duct*. Lymph from about three quarters of the body (see Figure 5-13) eventually drains into the thoracic duct and from it goes back into the blood. The thoracic duct joins the left subclavian vein at the angle where the internal jugular vein also joins it. (See Figure 5-8.) Lymph from the rest of the body drains into the right lymphatic ducts then on into the right subclavian vein.

Lymph nodes, or lymph glands as they are sometimes improperly called, are small oval structures located mainly in clusters along lymphatics (Figure 5-13). The structure of the lymph nodes makes it possible for them to perform an important protective function. They filter out injurious particles such as bacteria, soot, and cancer

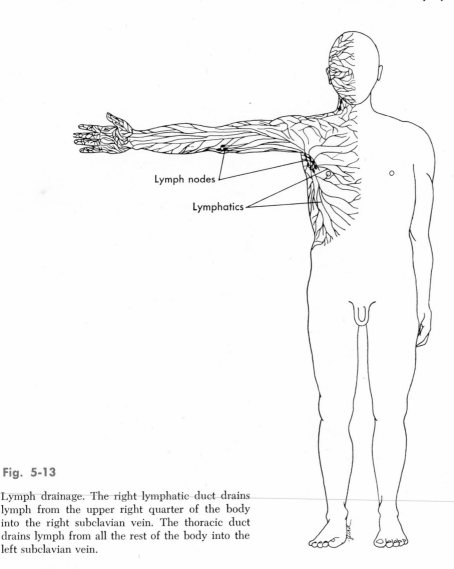

Lymph nodes

Lymphatics

Fig. 5-13

Lymph drainage. The right lymphatic duct drains lymph from the upper right quarter of the body into the right subclavian vein. The thoracic duct drains lymph from all the rest of the body into the left subclavian vein.

cells, thereby preventing them from entering the blood and circulating all over the body. Figure 5-14 illustrates lymph node structure and function. A surgeon uses his knowledge of lymph node function when he removes lymph nodes under the arms (axillary nodes) during an operation for breast cancer. These nodes may contain cancer cells filtered out of the lymph drained from the breast. A nurse, also, can use her knowledge of lymph node location and function; for example, when she takes care of a patient with an infected finger, she should watch the elbow and axillary regions for swelling and tenderness of the lymph nodes there; these nodes filter lymph returning from the hand and may become infected by the bacteria they trap. Lymph nodes also carry on another function, one mentioned earlier in this chapter. They form some of the white blood cells, namely lymphocytes and monocytes. (Lymphocyte functions are discussed on page 69.)

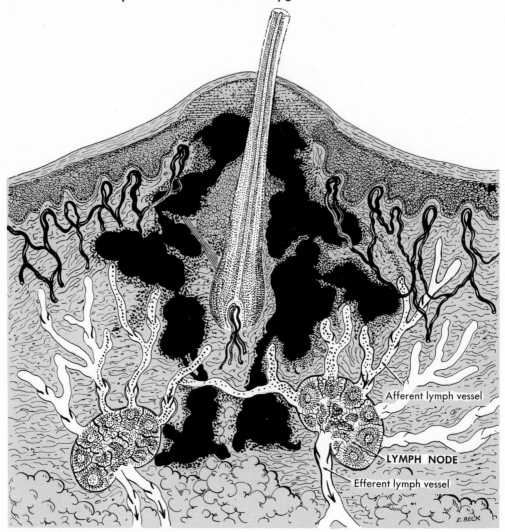

Fig. 5-14

Diagrammatic representation of lymph node structure and function. The drawing is of a skin section in which an infection surrounds a hair follicle. The black areas represent dead and dying cells (pus); black dots around the black areas represent bacteria. Leukocytes destroy many of the bacteria by phagocytosis, and as shown, many bacteria enter the lymph nodes by way of afferent lymphatics. The nodes filter and destroy most of the bacteria by enzymatic action.

THE SPLEEN

The spleen remains one of the mystery organs of the body. Quite a bit of doubt still exists about its functions. It is located in the upper left corner of the abdominal cavity just under the diaphragm. Except for blood vessels and nerves, it connects with no other organs.

The spleen's main functions seem to be to form lymphocytes and monocytes (as do the lymph nodes) and to act as the body's own blood bank. The spleen can store almost two cups of blood and quickly release it back into circulation when more blood is needed—during strenuous exercise and after hemorrhage, for instance.

Outline summary—The circulatory system

BLOOD STRUCTURE AND FUNCTIONS
Cells
1. Kinds
   a. Red cells (erythrocytes)
   b. White cells (leukocytes)
      1. Neutrophils
      2. Eosinophils
      3. Basophils
      4. Lymphocytes
      5. Monocytes
2. Numbers
   a. Red cells—4½ to 5 million per cubic millimeter of blood
   b. White cells—5,000 to 9,000 per cubic millimeter of blood
3. Formation
   a. Red bone marrow (myeloid tissue) forms all blood cells except lymphocytes and monocytes, which are formed by lymphatic tissue in lymph nodes and spleen
4. Functions
   a. Red cells—transport oxygen and carbon dioxide
   b. White cells—neutrophils and monocytes carry on phagocytosis; lymphocytes function to produce immunity
   c. Platelets—release prothrombinase; substance that starts blood clotting

Blood plasma
1. Definition—blood minus its cells
2. Composition—water containing many dissolved substances; for example, foods, salts, hormones
3. Amount of blood—varies with size and sex; 4 to 6 quarts about average
4. Reaction—slightly alkaline

Blood types (or blood groups)
1. Type A blood—type A antigens in red cells; anti-B type antibodies in plasma
2. Type B blood—type B antigens in red cells; anti-A type antibodies in plasma
3. Type AB blood—type A and type B antigens in red cells; no anti-A or anti-B type antibodies in plasma; therefore, type AB blood is called universal recipient blood
4. Type O blood—no type A or type B antigens in red cells; therefore, type O blood is called universal donor blood; both anti-A and anti-B type antibodies in plasma

5. Rh positive blood—Rh factor antigen present in red cells
6. Rh negative blood—no Rh factor present in red cells; no anti-Rh antibodies present naturally in plasma; anti-Rh antibodies, however, appear in plasma of Rh negative person if Rh positive blood cells have been introduced into his body

BLOOD CLOTTING
Summarized in Figure 5-2

THE HEART
1. Cavities—right atrium, right ventricle, left atrium, left ventricle
2. Wall—myocardium, composed of cardiac muscle
3. Lining—endocardium
4. Covering—pericardium
5. Valves—keep blood flowing in right direction through heart; prevent backflow
   a. Tricuspid—at opening of right atrium into ventricle
   b. Mitral (or bicuspid)—at opening of left atrium into ventricle
   c. Pulmonary semilunars—at beginning of pulmonary artery
   d. Aortic semilunars—at beginning of aorta

BLOOD VESSELS
1. Kinds
   a. Arteries—carry blood away from heart
   b. Veins—carry blood toward heart
   c. Capillaries—carry blood from arterioles to venules
2. Structure—see Figures 5-5 and 5-6
3. Names of main arteries—see Figure 5-7
4. Names of main veins—see Figure 5-8

CIRCULATION
1. Plan of circulation—see Figure 5-10
2. Portal circulation—detour of venous blood from stomach, pancreas, intestines, spleen through liver before return to heart
3. Plan of fetal circulation—see Figure 5-11

BLOOD PRESSURE
1. Blood pressure is push or force of blood in blood vessels
2. Highest in arteries, lowest in veins—see Figure 5-12
3. Blood pressure gradient causes blood to circulate—liquids can flow only from area where pressure is higher to where it is lower
4. Blood volume, heartbeat, and blood viscos-

ity are main factors that produce blood
pressure
5. Blood pressure varies within normal range
from time to time

PULSE
1. Definition—alternate expansion and recoil
of blood vessel wall
2. Places where you can count the pulse
easily—see page 82.

LYMPHATIC SYSTEM
1. Consists of lymphatic vessels, lymph nodes,
lymph
2. Lymph—the fluid in the lymphatic vessels;
lymph comes from blood by plasma filter-
ing out of blood capillaries to form interstitial fluid, some of which then enters
the lymph capillaries to become lymph and
be returned to blood by way of lymphatics;
largest lymphatic is thoracic duct—drains
lymph from all but upper right quarter
of body into left subclavian vein
3. Lymph nodes
   a. Located along certain lymphatics,
   usually in clusters; for example, at elbow, under arm, in groin, at knee
   b. Functions—filter out injurious particles
   such as microorganisms and cancer
   cells from lymph before it returns to
   blood; form some white blood cells
   (lymphocytes and monocytes)

SPLEEN
1. Forms some white blood cells (lymphocytes and monocytes)
2. Serves as blood bank for body—stores
blood until needed and then releases it
back into circulation

Review questions—The circulatory system

1. What is phagocytosis? What cells perform this function?
2. If someone asked you what blood plasma
is, what would you answer?
3. Suppose that you were asked what each
of the following terms means: erythrocytes, leukemia, leukocytosis, and throm-
bocytes. What would you answer?
4. Suppose that your doctor told you that
your "red count was 3 million." What
would that mean to you? What does "red
count" mean? Is 3 million normal for it?
Might the doctor say that you had any

of the following conditions—acidosis, anemia, leukopenia—with a red count of
this amount? Which one?
5. If you had appendicitis or some
other acute infection, would your white
count be more likely to be 2,000, 7,000,
or 15,000? Give a reason for your
answer.
6. Your circulatory system is the transportation system of your body. Mention some
of the substances it transports and tell
whether each is carried in blood cells or
in the blood plasma.
7. Do you think that advertisers' warnings to
"guard against acid blood" are justified?
Is there much danger of this?
8. Briefly explain what happens when blood
clots, including what makes it start to
clot.
9. You hear that a friend has a "coronary
thrombosis." What does this mean to
you?
10. Describe blood flow through the heart.
11. A patient has had an operation to repair
the mitral valve. Where is this valve,
and what is its function?
12. What are some differences between an
artery, a vein, and a capillary?
13. Considering that the function of the circulatory system is to transport substances
to and from the cells, do you think it is
true that, in one sense, capillaries are our
most important blood vessels? Give a reason for your answer.
14. The right ventricle of the heart pumps
blood to and through only one organ.
Which one?
15. What part of the heart pumps blood
through the systemic circulation, that is,
to and through all organs other than the
lungs?
16. All blood returns from the systemic circulation to what part of the heart?
17. What part of the heart pumps blood
through the pulmonary circulation, that
is, to and through the lungs?
18. Blood returning from the pulmonary circulation (from the lungs, in other words)
enters what part of the heart?
19. From which cavity of the heart does
blood rich in oxygen leave the heart to
be delivered to tissue capillaries all over
the body?
20. How do arterial blood and venous blood
differ with regard to their oxygen and
carbon dioxide contents?
21. Does every artery carry arterial blood and

every vein carry venous blood? If not, what exceptions are there?

22. Explain what is meant by "portal circulation."

23. Name the vein at the bend of the elbow into which substances are often injected and from which blood is sometimes withdrawn. (See Figure 5-9.)

24. Nurses frequently have to take a patient's blood pressure. Why is blood pressure important?

25. Sometimes a woman's arm becomes very swollen for a while after removal of a breast and the nearby lymph nodes and lymphatics, including some of those in the upper arm. Can you think of any reason why swelling occurs?

26. You could live without your spleen since it does not do anything vital for the body. What functions does it perform?

VI

The digestive system

You may already know quite a few facts about the digestive system. You probably know the names of the digestive organs, that some of them form a tube that extends through the body all the way from the mouth to the rectum, and that mucous membrane lines this tube. You most likely know what functions the digestive system performs—that it digests foods, absorbs the digested foods, and eliminates the wastes left over. Figure 1-3, page 6, lists the names of the organs that compose the digestive system. Figure 6-1, page 89, shows their locations.

In this chapter we shall discuss first the structure and function of individual digestive organs, then the functions of the digestive system as a whole, and finally we shall consider the complex and highly important process of metabolism.

THE STOMACH

The stomach is a kind of pouch that food enters after it has been chewed and swallowed and has passed through the esophagus. The stomach looks small after it is emptied, not much bigger than a large sausage, but it expands considerably after a large meal. Have you ever felt so uncomfortably full after eating that you could not take a deep breath? If so, it probably meant that your stomach was so full of food that it occupied more space than usual and pushed up against the diaphragm. This made it hard for the diaphragm to contract and move downward as much as necessary for a deep breath.

Three layers of smooth muscle, with fibers running lengthwise, around, and obliquely in the stomach wall, make the stomach one of the strongest internal or-

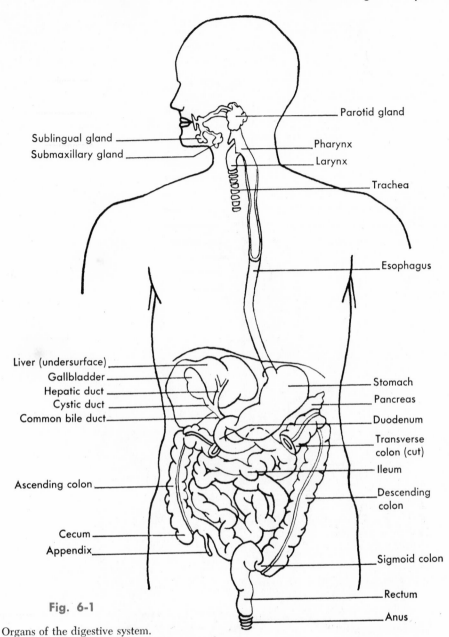

Fig. 6-1

Organs of the digestive system.

gans—well able to break up food into tiny particles and to mix them thoroughly with the gastric juice. Stomach muscle contractions take part in producing *peristalsis*, the movement that propels food down the length of the digestive tract. A mucous membrane lines the stomach. It contains thousands of microscopic glands that secrete gastric juice and hydrochloric acid into the stomach. When the stomach is empty, its lining lies in folds called *rugae*.

The lower part of the stomach is called the *pylorus*. It is a narrow section that joins the first part of the small intestine (the *duodenum*). Food is held in the stomach by the pyloric sphincter muscle long enough for partial digestion. The sphincter consists of circular smooth muscle fibers that stay contracted most of the time and thereby close off the opening of the pylorus into the duodenum. The fibers relax at intervals when part of the food is ready to leave the stomach, but sometimes they go into a spasm and do not relax normally; this condition is referred to as *pylorospasm*. It occurs in babies fairly often.

THE SMALL INTESTINE

The small intestine seems to be misnamed if we notice only its length—it is about 20 feet long; however, it is noticeably smaller around than the large intestine so that

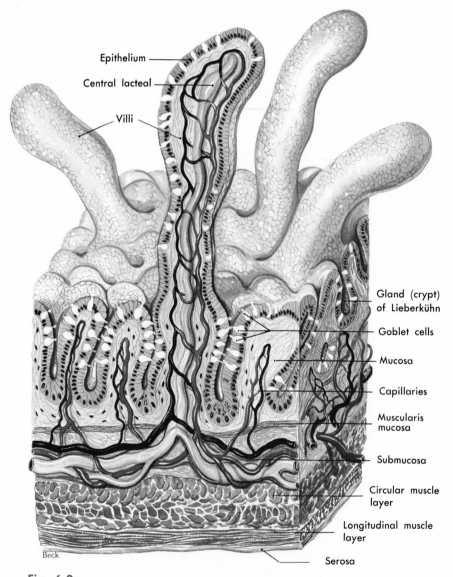

Fig. 6-2

Section of the intestinal mucosa showing villi and various microscopic structures.

in this respect its name is appropriate. Different names identify different sections of the small intestine. In the order in which food passes through them, they are the *duodenum, jejunum,* and *ileum.*

The mucous lining of the small intestine, like that of the stomach, contains thousands of microscopic glands. They are called *intestinal glands,* and they secrete the intestinal digestive juice. We should mention, also, something about the structure of the lining of the small intestine that makes it especially well suited to its function of food and water absorption. It is not perfectly smooth as it appears to be to the naked eye. Instead it has thousands of tiny "fingers," called *villi,* which, under the microscope, can be seen projecting into the hollow interior of the intestine. Inside each of these villi lies a rich network of blood and lymph capillaries. Turn to page 90 and look at Figure 6-2. Hundreds and hundreds of villi jut inward from the mucous lining. Imagine the lining perfectly smooth without any villi; think how much less surface area there would be for contact between capillaries and intestinal lining. Consider what an advantage a large contact area offers for faster absorption of food from the intestine into the blood and lymph—one more illustration of the prinicple that structure determines function.

Smooth muscle in the wall of the small intestine contracts to produce peristalsis, the wormlike movements which move food through the tract.

THE LARGE INTESTINE

The large intestine forms the lower part of the digestive tract. Its divisions—cecum, ascending colon, transverse colon, descending colon, sigmoid colon, and rectum—are shown in Figure 6-1. The rectum is about 7 or 8 inches long. Its external opening is called the *anus.* Two sphincter muscles stay contracted to keep the anus closed except during defecation. Smooth or involuntary muscle composes the inner anal sphincter,

but striated or voluntary muscle composes the outer one. This anatomical fact sometimes becomes highly important from a practical standpoint. For example, often after a person has had a stroke, the voluntary anal sphincter becomes paralyzed. This means, of course, that the individual will have no control over bowel movements. Or, in hospital language, the patient has "involuntary defecations."

ACCESSORY DIGESTIVE ORGANS

By accessory digestive organs we mean organs that are associated with the digestive tract but that do not form part of the tract itself. Incidentally, the terms "alimentary canal" and "gastrointestinal tract" (or G.I. tract for short) mean the same thing as the term "digestive tract"—namely, the mouth, pharynx, esophagus, stomach, and intestines. Accessory digestive organs, on the other hand, are the teeth, tongue, salivary glands, pancreas, liver, gallbladder, and vermiform appendix.

The teeth

By the time a baby is 2 years old, he probably has his full set of 20 baby teeth. By the time a young adult is somewhere between 17 and 24 years old, he usually has his full set of 32 permanent teeth. The average age for cutting the first tooth is about 6 months, and the average age for losing the first baby tooth and starting to cut the permanent teeth is about 6 years. Figure 6-3 gives the names of the teeth and shows which ones are lacking in the deciduous or baby set. Figure 6-4 illustrates tooth structure.

The salivary glands

The *parotid glands* lie just below and in front of each ear at the angle of the jaw—an interesting anatomical fact because it explains why people who have mumps (infection of the parotid gland) often complain that it hurts when they open their mouths or chew; these movements squeeze

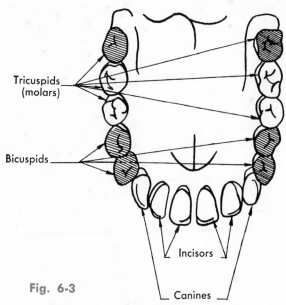

Fig. 6-3

Teeth. The permanent set of teeth includes all of those shown. (Both the upper and lower jaws have the same number and arrangement of teeth.) The deciduous set or "baby teeth" lacks the teeth striped in the diagram.

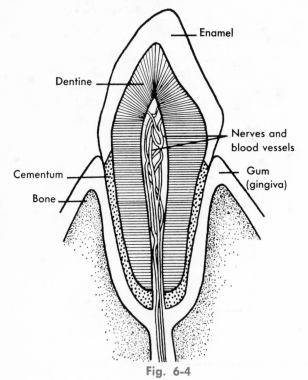

Fig. 6-4

Longitudinal section of an incisor tooth.

the tender inflamed gland. To see the openings of the ducts of the parotid glands, look on the insides of your cheeks opposite the second tricuspid tooth, on either side of the upper jaw. Besides the parotid glands there are also two other pairs of salivary glands—the *submaxillary* and the *sublingual glands*. Their ducts open into the floor of the mouth.

The liver and gallbladder

The liver is the largest gland in the body and one of the most important. It fills the upper right section of the abdominal cavity, and even extends part way over onto the left side. One of its many functions is to secrete *bile*. This digestive juice drains out of the liver by way of the *hepatic ducts*. Between meals, bile goes up the cystic duct into the gallbladder (located on the undersurface of the liver) for concentration and storage. After meals, when fats enter the duodenum, the gallbladder ejects bile into the cystic duct down the common bile duct into the duodenum. Blood normally contains some bile pigments; however, if it contains too much, the skin looks yellow and we say that the person has *jaundice*. One thing that can cause jaundice is the absorption of excessive amounts of bile into the blood, due to a blocking off of one or more of the ducts that drain bile out of the liver into the intestine. Examine Figure 6-5. If a gallstone were to block the cystic duct, would this prevent drainage of bile from the liver into the intestine, or would it merely prevent bile from entering or leaving the gallbladder? A gallstone or any other obstruction of either the hepatic ducts or the common bile duct will prevent bile from draining from the liver into the intestine and therefore will produce jaundice, but obstruction of the cystic duct will do neither.

In addition to secreting bile, liver cells perform other functions necessary for healthy survival. Two of the most important are these: they help keep the amount of

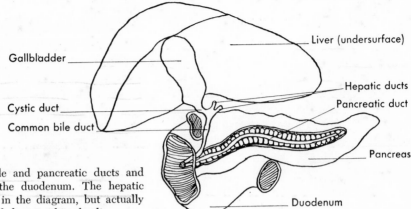

Gallbladder

Liver (undersurface)

Hepatic ducts

Pancreatic duct

Cystic duct

Common bile duct

Pancreas

Duodenum

Fig. 6-5

Diagram showing bile and pancreatic ducts and their openings into the duodenum. The hepatic ducts are shown cut in the diagram, but actually they are extensions of ducts within the liver.

sugar in the blood at a normal level and they make various blood proteins. When blood sugar (mainly glucose) starts to increase, liver cells change some of it to glycogen and store this compound. But when blood sugar starts to decrease, they change the glycogen back to glucose and release it into the blood. Normally liver cells are busy after meals changing glucose to glycogen. *Glycogenesis* is the name of this process. It prevents blood sugar from increasing to a dangerously high level when it is being absorbed from the intestine into the blood. Later in the between-meal period, when glucose absorption has ended, liver cells change glycogen back to glucose (*glycogenolysis*) and release the glucose into the blood. By this process they guard against a dangerous decrease in blood sugar when glucose is leaving blood for use by body cells but not entering it from the intestine.

Liver cells synthesize several kinds of protein compounds which they release into the blood, where these compounds are called the "blood proteins" or "plasma proteins." Prothrombin and fibrinogen are the names of two of the blood proteins formed by liver cells. What do you remember about the function of these substances? (See page 71.)

The pancreas

The pancreas lies behind the stomach. Some of its cells secrete *pancreatic juice* into ducts. This most important digestive juice drains out of the *pancreatic duct* into the duodenum at the same place that the bile enters. Some other cells located in the pancreas secrete *insulin* into the blood instead of into ducts. In other words, these cells are ductless or endocrine glands. Because they are separate from the cells that secrete pancreatic juice, they are called islands—*islands of Langerhans.* Insulin, as you probably know, is a hormone absolutely vital for maintaining a normal amount of blood sugar. (See p. 96.)

The appendix

The appendix is a dead-end tube off the cecum. Although it serves no known function, it is often a nuisance when its mucous lining becomes inflamed—the well-known affliction, appendicitis.

DIGESTION

All of the digestive organs work together to perform the function known as digestion. Digestion is the process that prepares foods for absorption; that is, it changes foods into substances that can pass through the mucous membrane lining of the small intestine into the blood and eventually into cells. Digestion consists of several mechanical processes and many chemical reactions. The mechanical processes—chewing, swallowing, and peristalsis—break the food into tiny particles, mix them with the digestive juices, and move them along the digestive tract. Molecules that compose the foods that we eat are too large for absorption. Digestion changes this. Chemical reactions catalyzed by the *digestive enzymes,* break down the large nonabsorbable food molecules into smaller molecules that can readily

pass through the intestinal mucosa into blood or lymphatic capillaries.

Protein digestion starts in the stomach. Two enzymes (rennin and pepsin) in the gastric juice cause the giant protein molecules to break up into somewhat simpler compounds. Then, in the intestine, other enzymes finish the job of protein digestion. Pancreatic juice contains one protein-digesting enzyme (trypsin), and intestinal juice contains another (erepsin). Every protein molecule is made up of many amino acids joined together. When enzymes

Tabel 6-1. Chemical digestion

Digestive juices and enzymes	Enzyme digests (or hydrolyzes)	Resulting product*
Saliva Amylase (ptyalin)	Starch (polysaccharide or complex sugar)	Maltose (a disaccharide or double sugar)

Gastric juice
  Protease (pepsin) plus hydrochloric acid
  Lipase (of little importance)
	Proteins, including casein
	
	Emulsified fats (butter, cream, etc.)
		Proteoses and peptones (partially digested proteins)
		Fatty acids and glycerol

Bile contains no enzymes	Large fat droplets (unemulsified fats)	Small fat droplets or emulsified fats

Pancreatic juice
  Protease (trypsin)†
 	
  Lipase (steapsin)
  Amylase (amylopsin)
	Proteins (either intact or partially digested)
	Bile–emulsified fats
	Starch
		Proteoses, peptides, and *amino acids*
		Fatty acids and glycerol
		Maltose

Intestinal juice (succus entericus)
  Peptidases
  Sucrase
 	
  Lactase
 	
  Maltase
	Peptides
	Sucrose (cane sugar)
	
	Lactose (milk sugar)
	
	Maltose (malt sugar)
		Amino acids
		Glucose and fructose‡ (simple sugars or monosaccharides)
		Glucose and galactose (simple sugars)
		Glucose (grape sugar)

*Substances in italics are end products of digestion or, in other words, completely digested foods ready for absorption.
†Secreted in inactive form (trypsinogen); activated by enterokinase, an enzyme in the intestinal juice.
‡Glucose is also called dextrose; fructose is called levulose.

have split up the large protein molecule into its separate amino acids, protein digestion is completed. Hence, the "end product of protein digestion" is amino acids. For obvious reasons, amino acids are also referred to as "protein building blocks."

Very little carbohydrate digestion occurs before food reaches the small intestine. (Starches and sugars are carbohydrates.) Gastric juice contains no enzymes that act on carbohydrates, and those in saliva usually have too little time to do their work because so many of us swallow our food so fast. But once the food reaches the small intestine, pancreatic and intestinal juice enzymes digest the starches and sugars. A pancreatic enzyme (amylopsin or amylase) starts the process by changing starches into maltose. Three intestinal enzymes digest sugars, changing them into glucose (also called dextrose). Maltase digests maltose (malt sugar), sucrase digests sucrose (ordinary cane sugar), and lactase digests lactose (milk sugar). The end product of carbohydrate digestion is mainly glucose, a simple sugar.

Not only very little carbohydrate digestion but also very little fat digestion occurs before food reaches the small intestine. An enzyme in gastric juice (gastric lipase) digests some fat in the stomach, but most fats go undigested until after bile emulsifies them—that is, breaks the fat droplets into very small droplets. After this takes place, the pancreatic enzyme (steapsin or pancreatic lipase) splits up the fat molecules into fatty acids and glycerol (glycerin). The end products of fat digestion, then, are fatty acids and glycerol. Have you noticed that only one digestive juice—pancreatic juice—can digest all three kinds of foods? In what organ does the pancreatic juice do its work?

Table 6-1 summarizes the main facts about chemical digestion. Enzyme names indicate the type of food digested by the enzyme. For example, the name *amylase*

indicates that the enzyme digests carbohydrates (starches and sugars). The name *protease* indicates a protein-digesting enzyme, and the name *lipase* means a fat-digesting enzyme. When carbohydrate digestion has been completed, starches and complex sugars (polysaccharides) have been changed mainly to glucose, a simple sugar or monosaccharide. The end products of protein digestion, on the other hand, are amino acids. Fatty acid and glycerol are the end products of fat digestion. Use information in Table 6-1 to answer questions 7 to 13, page 99.

ABSORPTION

When we talk about food being absorbed, we mean that after it is digested it moves through the mucous membrane lining of the small intestine into the blood and lymph. Or, stated in another way, food absorption means that molecules of amino acids, glucose, fatty acids, and glycerol go from the inside of the intestines into the circulating fluids of the body. Absorption of foods is just as essential a process as the digestion of foods. The reason is fairly obvious. As long as food stays in the intestines, it cannot nourish the millions of cells composing all other parts of the body. Their lives depend upon the absorption of digested food and its transportation to them by the circulating blood. On page 91 we mentioned something about the structure of the intestinal lining that enables it to absorb foods rapidly. Do you recall what this is?

METABOLISM

A good phrase to remember in connection with the word metabolism is "the use of foods," for basically this is what metabolism is—the use the body makes of foods once they have been digested, absorbed, and circulated to cells. What use does the body make of foods? It uses them in two ways: as an energy source, and as building blocks for making complex chemical compounds. Before they can be used in these two ways, foods have to enter cells and there undergo many chemical changes. All chemical reactions that release energy from food molecules together make up the process of catabolism—a vital process since it is the only way that the body has of supplying itself with energy for doing any of its many kinds of work. The many chemical reactions which build food molecules into more complex chemical compounds together constitute the process of anabolism. Catabolism and anabolism together make up the process of metabolism. Reread the discussion of metabolism on pages 9 to 10. Reexamine Figure 1-5, page 10.

Something worth noticing is that the amount of food in the blood normally does not change very much, not even when we go without food for many hours, or when we exercise and use a lot of food for energy, or when we sleep and use little food for energy. The amount of glucose, for example, usually stays at about 80 to 120 milligrams in 100 milliliters of blood.

Several hormones help regulate carbohydrate metabolism so as to keep blood sugar normal. *Insulin* is one of the most important of these. It acts in some way not yet definitely known to make glucose leave the blood and enter the cells at a more rapid rate. As insulin secretion increases, more glucose leaves the blood and enters the cells. The amount of glucose in the blood therefore tends to decrease while the rate of glucose metabolism in cells tends to increase. Too little insulin secretion, such as occurs in diabetes mellitus, produces the opposite effects. Less glucose leaves the blood and enters the cells, therefore, more glucose remains in the blood and less glucose is metabolized by cells. In other words, high blood sugar (hyperglycemia) and a low rate of glucose metabolism characterize insulin deficiency. Insulin is the only hormone which functions to lower the blood sugar level. Several hormones, on the other hand, tend to increase it. Growth

hormone secreted by the anterior pituitary gland, hydrocortisone secreted by the adrenal cortex, and epinephrine secreted by the adrenal medulla are three of the most important hormones that tend to increase blood sugar. More information about these hormones and others that help control metabolism appears on pages 133 to 137.

The carbohydrate glucose is the body's preferred energy food. Human cells seem to prefer to catabolize glucose rather than other substances, and they do so as long as enough glucose enters them to supply their energy needs. They also, however, anabolize small amounts of carbohydrates. For example, the process of glycogenesis, described on page 93 is an important kind of carbohydrate anabolism.

Fats, like carbohydrates, are primarily energy foods. If cells have inadequate amounts of glucose to catabolize, they immediately shift to the catabolism of fats for their energy supply. This happens normally whenever a person goes without food for many hours. It happens abnormally in diabetic individuals. Because of an insulin deficiency, too little glucose enters the cells of a diabetic person to supply all of his energy needs, so that the cells catabolize fats to make up the difference. In all persons, fats not needed for catabolism are anabolized and stored in adipose tissue.

Protein catabolism occurs to some extent, but more important is protein anabolism, the process by which the body builds amino acids into complex protein compounds—for example, enzymes and proteins that form the structural parts of the cell.

Metabolic rates

Every nurse, sooner or later, hears the letters B.M.R. They stand for *basal metabolic rate*. Suppose you were to "have a B.M.R." You would breathe through your mouth into a tube connected to a machine, which would measure the amount of oxygen you breathe. You would be lying down

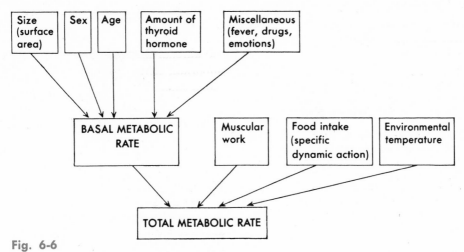

Fig. 6-6

Factors that determine the basal and the total metabolic rates.

but awake. You would not have eaten for 12 hours or more, and the room would be comfortably warm. (Those last two sentences describe basal conditions.) From the amount of oxygen you had breathed in, the technician could compute the amount of food your body had catabolized under the basal conditions—could compute your B.M.R., in other words. She might report this as so many calories (that is, so much heat from food catabolized). More often she reports it simply as "normal" or a certain percentage above or below normal. Your normal B.M.R. might be 1,200 calories. This would not be normal for everyone, however. Someone bigger or younger than you might have a normal B.M.R. of 1,400 calories. Someone smaller or someone the same size but older, on the other hand, might have a normal basal rate of only 1,000 calories. A man normally has a higher basal rate than a woman his same size and age.

A person's B.M.R. represents the amount of food that his body must catabolize each day for him simply to stay alive and awake in a comfortably warm environment. To provide energy for his doing any muscular work and for digesting and absorbing any food, an additional amount of food must be catabolized. How much more depends upon how much work the individual does. The more active he is, the more food his body must catabolize and the higher his total metabolic rate will be. The *total metabolic rate* or T.M.R. is the total amount of energy (expressed in calories) used by the body per day (see Figure 6-6).

When a person's total metabolic rate equals the total calories in all the food that he eats per day, that person's weight remains constant (except for possible variations resulting from water retention or water loss). When his daily food contains more calories than his total metabolic rate, he gains weight; when his daily food contains fewer calories than his total metabolic rate, he loses weight. These weight control principles never fail to operate. Nature never forgets to count calories. Reducing diets make use of this knowledge. They contain fewer calories than the total metabolic rate of the individual eating the diet.

Outline summary—The digestive system

STOMACH

1. Size—expands after large meal; about size of large sausage when empty
2. Pylorus—lower part of stomach; pyloric sphincter muscle closes opening of pylorus into duodenum
3. Wall—many smooth muscle fibers; contractions produce churning movements and peristalsis
4. Lining—mucous membrane; many microscopic glands that secrete gastric juice and hydrochloric acid into stomach; mucous membrane lies in folds (rugae) when stomach is empty

SMALL INTESTINE

1. Size—about 20 feet long but only an inch or so in diameter
2. Divisions
   a. Duodenum
   b. Jejunum
   c. Ileum
3. Wall—contains involuntary muscle fibers that contract to produce peristalsis
4. Lining—mucous membrane; many microscopic glands (intestinal glands) secrete intestinal juice; villi (microscopic finger-shaped projections from surface of mucosa into intestinal cavity) contain blood and lymph capillaries

LARGE INTESTINE

1. Divisions
   a. Cecum
   b. Colon—ascending, transverse, descending, and sigmoid
   c. Rectum
2. Opening to exterior—anus
3. Wall—contains smooth muscle fibers that contract to produce churning, peristalsis, and defecation
4. Lining—mucous membrane

ACCESSORY DIGESTIVE ORGANS

The teeth

1. Twenty teeth in temporary set; average age for cutting first tooth about 6 months; set complete at about 2 years of age
2. Thirty-two teeth in permanent set; 6 years about average age for starting to cut first permanent tooth; set complete usually between ages of 17 to 24 years
3. Names of teeth—see Figure 6-3
4. Structures of tooth—see Figure 6-4

The salivary glands

1. Parotid (LOCATED NEAR EAR)
2. Submaxillary
3. Sublingual

The liver and gallbladder

1. Size and location—largest gland; fills upper right section of abdominal cavity and extends over into left side
2. Functions—secretes bile, helps maintain normal blood sugar by carrying on processes of glycogenesis and glycogenolysis; forms prothrombin, fibrinogen, and certain other blood proteins; also performs several other functions
3. Ducts
   a. Hepatic—drains bile from liver
   b. Cystic—duct by which bile enters and leaves gallbladder
   c. Common bile—formed by union of hepatic and cystic ducts; drains bile from hepatic or cystic ducts into duodenum

The pancreas

1. Location—behind stomach
2. Functions
   a. Pancreatic cells secrete pancreatic juice into pancreatic ducts; main duct empties into duodenum
   b. Islands of Langerhans—cells not connected with pancreatic ducts; secrete insulin into blood

The appendix

Blind tube off cecum; no known function

DIGESTION

Meaning—changing foods so that they can be absorbed and used by cells

Mechanical digestion

Chewing, swallowing, and peristalsis break food into tiny particles, mix them well with the digestive juices, and move them along the digestive tract

Chemical digestion

Breaks up large food molecules into compounds having smaller molecules; brought about by digestive enzymes

Protein digestion

Starts in stomach; completed in small intestine

1. Gastric juice enzymes, rennin and pepsin, partially digest proteins
2. Pancreatic enzyme, typsin, completes digestion of proteins to amino acids

3. Intestinal enzyme, erepsin, completes digestion of partially digested proteins to amino acids

Carbohydrate digestion

Mainly in small intestine.
1. Pancreatic enzyme, amylopsin, changes starches to maltose
2. Intestinal juice enzymes
   a. Maltase changes maltose to glucose
   b. Sucrase changes sucrose to glucose
   c. Lactase changes lactose to glucose

Fat digestion

1. Gastric lipase changes small amount of fat to fatty acids and glycerin in stomach
2. Bile contains no enzymes but emulsifies fats (breaks fat droplets into very small droplets)
3. Pancreatic lipase changes emulsified fats to fatty acids and glycerin in small intestine

ABSORPTION

1. Meaning—digested food moves from intestine into blood or lymph
2. Where absorption occurs—foods and water from small intestine; water also absorbed from large intestine

METABOLISM

1. Meaning—use of foods by body cells for energy and building complex compounds
2. Catabolism—breaks food molecules down into carbon dioxide and water, releasing their stored energy; oxygen used up in catabolism
3. Anabolism—builds food molecules into complex substances
4. Carbohydrates primarily catabolized for energy but small amounts are anabolized by glycogenesis (a series of chemical reactions which changes glucose to glycogen —occurs mainly in liver cells where glycogen is stored); glycogenolysis is process (series of chemical reactions) by which glycogen is changed back to glucose
5. Blood sugar (or blood glucose)—normally stays between about 80 and 120 milligrams per 100 milliliters of blood; *insulin* accelerates movement of glucose out of blood into cells; therefore tends to decrease blood glucose and increase glucose catabolism
6. Fats both catabolized to yield energy and anabolized to form adipose tissue
7. Proteins primarily anabolized and secondarily catabolized

Metabolic rates

1. Basal metabolic rate (B.M.R.)—rate of metabolism when person is lying down, but awake, when about 12 hours have passed since last meal, and when environment is comfortably warm
2. Total metabolic rate (T.M.R.)—the total amount of energy, expressed in calories, used by the body per day

Review questions—The digestive system

1. What organs form the gastrointestinal tract?
2. Identify each of the following structures: jejunum, cecum, colon, duodenum, and ileum.
3. If you inserted 9 inches of an enema tube through the anus, the tip of the tube would probably be in what structure?
4. How many teeth should an adult have?
5. How many teeth should a child 2½ years old have? Would he have some of each of the following teeth: incisors, canines, bicuspids, and tricuspids?
6. Identify each of the following:
   rugae   islands of Langerhans
   pylorus   parotid glands
   villi
7. In what organ does the digestion of starches begin?
8. What digestive juice contains no enzymes?
9. Only one digestive juice contains enzymes for digesting all three kinds of food. Which juice is this?
10. Which kind of food is not digested in the stomach?
11. Which digestive juice emulsifies fats?
12. What three digestive juices act on foods in the small intestine?
13. What juices digest carbohydrates? proteins? fats?
14. Explain as briefly and clearly as you can what each of the following terms means:
    digestion   anabolism
    absorption   catabolism
    metabolism
15. Explain why you think the following statement is true or false: "If you do not want to gain or lose weight but just stay the same, you must eat just enough food to supply the calories of your B.M.R. If you eat more than this, you will gain; if you eat less than this, you will lose."

VII

The respiratory system

No one needs to be told how important his respiratory system is. Everyone knows this intuitively. For example, think how panicky you would feel if suddenly you could not breathe for a few seconds. Of all the substances that cells and therefore the body as a whole must have to survive, oxygen is by far the most crucial. A person can live a few weeks without food, a few days without water, but only a few minutes without oxygen. The respiratory system serves the body much as a lifeline to an oxygen tank serves a deep-sea diver. In this chapter the structural plan of the respiratory system will be considered first, then the respiratory organs will be discussed individually, and finally some facts about respiration which are important for a nurse to know will be given.

STRUCTURAL PLAN

The respiratory organs are the nose, pharynx, larynx, trachea, bronchi, and lungs. Their basic plan is that of a tube with many branches ending in millions of extremely tiny, extremely thin-walled sacs called *alveoli*. A network of capillaries fits like a tight-fitting hairnet around each microscopic alveolus. Incidentally, this is a good place for us to think again about an idea or principle already mentioned several times, namely, that structure determines function. The function of alveoli—in fact, the function of the entire respiratory system—is to bring air close enough to blood for oxygen to get into the blood and carbon dioxide to get out of it. Two facts about the structure of alveoli make them able to perform this function admirably. For one thing, the wall of each alveolus is made up of a single layer of cells and so are the walls of the capillaries around it. This means that between the blood in the capillaries and the air in the alveolus there is a barrier probably less than 1/5000 of an inch thick! The other advantage of alveoli

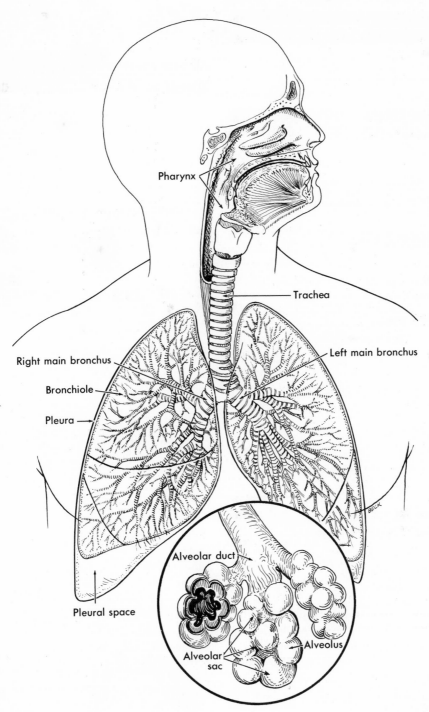

Fig. 7-1

Structural plan of the respiratory organs showing the pharynx, trachea, bronchi, and lungs. The inset shows the grapelike alveolar sacs where the interchange of oxygen and carbon dioxide takes place through the thin walls of the alveoli. Capillaries (not shown) surround the alveoli.

for accomplishing their function is the fact that there are so many of them—literally millions. This means that together they make an enormous surface (in the neighborhood of 1,100 square feet, an area many times larger than the surface of the entire body) where large amounts of oxygen and carbon dioxide can rapidly be exchanged.

THE NOSE

Air enters the respiratory tube through the nostrils (nares) into the right and left nasal cavities. A partition, the nasal septum, separates these two cavities; mucous membrane lines them. The surface of the nasal cavities is moist from mucus and warm from blood flowing just under it. Nerve endings responsible for the sense of smell are located in the nasal mucosa. Four pairs of sinuses (frontal, maxillary, sphenoidal and ethmoidal) drain into the nasal cavities. The nasal mucosa also lines the sinuses, and since it becomes infected whenever we have a cold, sinus infection often develops from colds.

THE PHARYNX

The pharynx is what most of us call the throat. It serves the same purpose for the respiratory and digestive tracts as a hallway serves for a house. Both air and food pass through the pharynx on their way to the lungs and the stomach, respectively. Air enters the pharynx from the two nasal cavities and leaves it by way of the larynx; food enters it from the mouth and leaves it by way of the esophagus. In addition, a pair of tubes opens into the pharynx—the eustachian (auditory) tubes that lead from each middle ear into the pharynx. Two pairs of organs that seem to give more trouble than service to the body (namely, the tonsils and the adenoids) are also located in the pharynx.

THE LARYNX

The larynx or voice box is located just below the pharynx. It is composed of sev-

eral pieces of cartilage. You know the largest of these (the thyroid cartilage) as the "Adam's apple." One of the cartilages of the larynx, the epiglottis, acts as a lid to close off the larynx when we swallow. Occasionally this lid does not work properly; then we cough and choke and say that we have "swallowed down our Sunday throat" —which means that food or liquid has entered the larynx where only air should go.

Two short fibrous bands called the vocal cords stretch across the interior of the larynx. Muscles that attach to the cartilages of the larynx can pull on these cords in such a way that they become either tense and short or long and relaxed. When they are short and tense, the voice sounds high pitched; when they are long and relaxed, it sounds low pitched.

THE TRACHEA

By pushing with your fingers against your throat about an inch above the breast bone, you can feel the shape of the trachea or windpipe. Only if you use considerable force can you squeeze it closed. Nature has taken precautions to keep this lifeline open. Its framework is made of an almost noncollapsible material—namely, 15 or 20 C-shaped rings of cartilage placed one above the other with only a little soft tissue between them. Despite this structural safeguard, closing off of the trachea does sometimes occur. A tumor or an infection may enlarge the lymph nodes of the neck so much that they squeeze the trachea shut, or a person may aspirate (breathe in) a piece of food or something else that blocks the windpipe. Since air has no other way to get to the lungs, complete tracheal obstruction causes death in a matter of minutes.

THE BRONCHI, BRONCHIOLES, AND ALVEOLI

One way to picture the thousands of air tubes that make up the lungs is to think of an upside-down tree. The trachea is the main trunk of this tree; the right bronchus (the tube leading into the right lung) and the left bronchus (the tube leading into the left lung) are the trachea's first branches. In each lung they branch into smaller bronchi, which branch into bronchioles. The smallest bronchioles end in structures shaped like miniature bunches of grapes (Figure 7-1). The smallest bronchioles subdivide into microscopic-sized tubes called *alveolar ducts,* which resemble the main stem of a bunch of grapes. Each alveolar duct ends in several *alveolar sacs,* each of which resembles a cluster of grapes, and the wall of each alveolar sac is made up of numerous *alveoli,* each of which resembles a single grape. How well the structure of the alveoli suits them to their function was discussed on pages 100 and 102.

Because the air tubes of the bronchial tree are imbedded in connective tissue, you cannot see them by looking at the lungs from the outside. But if you cut into the lungs, the bronchi and bronchioles show up clearly.

THE LUNGS

The lungs are fairly large organs. They fill the entire chest cavity (all but the space in the center occupied mainly by the heart and large blood vessels). The narrow part of each lung, up under the collarbone, is its *apex;* the broad lower part, resting on the diaphragm, is its *base.* A thin layer of moist slippery membrane, the *pleura,* covers the outer surface of the lungs and lines the inner surface of the rib cage. When the lungs move in breathing, the two layers of pleura glide smoothly against each other. (Do you recall another place with this same structural arrangement that also permits friction-free movement of a vital organ? Check your answer on page 72.)

In the condition called pleurisy, the pleura becomes inflamed and respirations become difficult and painful.

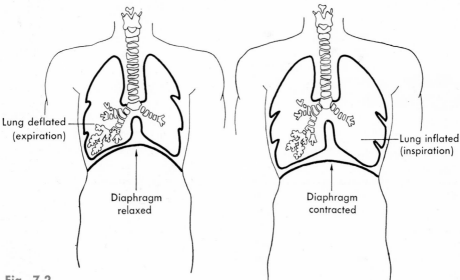

Lung deflated (expiration)

Lung inflated (inspiration)

Diaphragm relaxed

Diaphragm contracted

Fig. 7-2

Respiration. Note the difference in size of the thoracic cavity and lungs during expiration and inspiration. Contraction of the diaphragm and chest-elevating muscles enlarges the thoracic cavity and lungs in all three dimensions. This lowers the pressure within the air passages of the lungs and causes outside air to move down them into the alveoli.

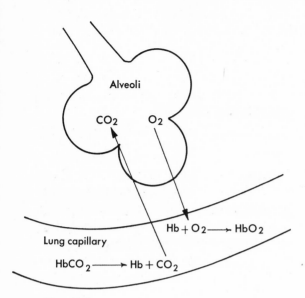

Fig. 7-3

Diagram showing the exchange of gases between blood and air in the lungs. As venous blood flows through the lung capillaries, the loss of carbon dioxide and the gain of oxygen change it to arterial blood.

RESPIRATION

Respiration means exchange of gases (oxygen and carbon dioxide) between a living organism and its environment. If the organism consists of only one cell, gases can move directly between it and the environment. If, however, the organism consists of billions of cells, as do our bodies, then most of its cells are too far from the air for a direct exchange of gases. To overcome this difficulty, an organ—the lungs—is provided where air and a circulating fluid (blood) can come close enough to each other for oxygen to move out of the air into blood while carbon dioxide moves out of the blood into air. This exchange of gases between air and blood is what we call breathing or respiration. The movement of gases between the blood and cells, however, is also respiration—cellular respiration or cell breathing.

Exchange of gases in the lungs

As blood flows through the thousands of tiny lung capillaries, carbon dioxide leaves it and oxygen enters it. This double exchange of gases between the blood in the lung capillaries and the air in the alveoli comes about in the following way. Venous blood enters lung capillaries. (These are the only capillaries where this is true; arterial blood enters all other capillaries— "tissue capillaries" as they are called. Do you recall why venous blood enters lung capillaries? Where has it come from? See pages 77 to 78 if you are not sure.) The hemoglobin inside many of the red blood cells in venous blood is combined with carbon dioxide rather than with oxygen, and so is called carbohemoglobin or carbamino-hemoglobin. As the blood flows along through the lung capillaries, carbohemoglobin breaks down into carbon dioxide and hemoglobin. The hemoglobin then picks up oxygen molecules while carbon dioxide molecules move out of the blood. They pass quickly through the thin capillary-alveolar membrane into the alveoli. Then finally from the alveoli, carbon dioxide leaves the body in the expired air. Some carbon dioxide is also present in blood in the form of carbonic acid, formed by water combining with carbon dioxide as shown by the following equation:

$$H_2O + CO_2 \longrightarrow H_2CO_3 \text{ (carbonic acid)}$$

Some of the carbon dioxide that leaves the lung capillary blood has come from the reversing of the above equation, from carbonic acid breaking down again to become water and carbon dioxide. Therefore, as carbon dioxide leaves the blood in the lung capillaries, both the carbonic acid content of blood and its carbon dioxide content decrease.

The exchange of gases between lung capillary blood and the alveolar air (carbon dioxide out of the blood, oxygen into the blood) changes venous blood to arterial blood. Figure 7-3 shows these changes. In summary, the exchange of gases in the lung capillaries causes arterial blood to differ from venous blood in these ways: arterial blood contains more oxygen, but less carbon dioxide and less carbonic acid than does venous blood, and because it contains less carbonic acid, arterial blood is more alkaline than venous blood.

Imagine a patient who hyperventilates over a period of hours. (Hyperventilation means increased breathing; more air moving in and out of the lungs per minute.) More carbon dioxide than normal would leave this patient's blood every minute. How would this affect the carbonic acid content of his blood? Increase it or decrease it? Do you think, therefore, that prolonged hyperventilation would tend to make his blood more or less alkaline than normal? In other words, would you expect this patient to develop alkalosis or acidosis?

Exchange of gases in the tissues

The exchange of gases between the blood in tissue capillaries and the cells that make up the tissues is just the opposite of the exchange of gases between the blood in lung capillaries and the air in alveoli. As shown in Figure 7-4, in the tissue capillaries, oxyhemoglobin breaks down into oxygen and hemoglobin. Oxygen molecules move rapidly out of the blood through the tissue capillary membrane into the interstitial fluid and on into the cells that compose the tissue. While this is happening, carbon dioxide molecules are leaving the cells, entering the tissue capillaries, and uniting with hemoglobin molecules to form carbohemoglobin. In other words, the arterial blood that enters tissue capillaries becomes changed into venous blood as it flows through them.

Mechanics of breathing

Breathing involves not only the organs of the respiratory system but also the brain, spinal cord, nerves, certain skeletal muscles, and even some bones and joints. In breathing, nerve impulses stimulate the diaphragm to contract, and as it contracts its shape changes. The dome-like shape of the diaphragm flattens out so that it no longer protrudes into the chest cavity. As you can see in Figure 7-2, this flattening of the diaphragm makes the chest cavity longer from top to bottom. Other muscle contractions raise up the rib cage to make the chest cavity wider and greater in depth from front to back. As the chest cavity enlarges, the lungs expand along with it and

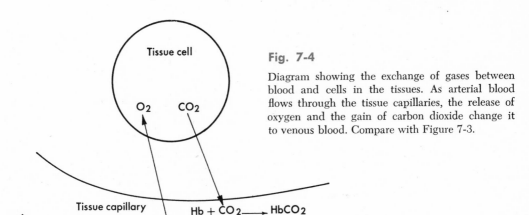

Fig. 7-4

Diagram showing the exchange of gases between blood and cells in the tissues. As arterial blood flows through the tissue capillaries, the release of oxygen and the gain of carbon dioxide change it to venous blood. Compare with Figure 7-3.

air rushes into them and down into the alveoli. This part of respiration is called inspiration. For expiration to take place, the diaphragm and other respiratory muscles relax, and the chest cavity becomes smaller, thereby squeezing air out of the lungs.

Ordinarily we take about a pint of air into our lungs. Because this amount comes and goes regularly like the tides of the sea, it is referred to as the *tidal air*. The largest amount of air that we can breathe in and out in one inspiration and expiration is known as the *vital capacity*; in most adults this is about 3½ quarts. A special device, the spirometer, is used to measure the amount of air exchanged in breathing. Physicians use this information mainly with patients who have lung or heart disease, because these conditions often lead to abnormal volumes of air being moved in and out of the lungs.

By now you know that the body uses oxygen to get energy for the work it has to do. Briefly, energy is stored in foods and the oxidation reactions of catabolism make this energy available for all kinds of cellular work. (See pages 9 to 10.) Therefore, the more work the body does, the more oxygen must be delivered to its millions of cells. One way this is accomplished is by increasing the rate and depth of respirations. Whereas we may take only 16 breaths a minute when we are not moving about, when we are exercising we take considerably more than this. Not only do we take more breaths; we also breathe in more air with each breath. Instead of about a pint of air, we may breathe deeply enough to take in several pints—sometimes even up to the limit of our vital capacity.

The way respirations are made to increase during exercise is an interesting example of the body's automatic regulation of its vital functions. During exercise, cells carry on catabolism at a faster rate than usual. This means that they form more carbon dioxide and that more carbon di-

oxide enters the blood in the tissue capillaries; therefore, the blood's carbon dioxide concentration starts to increase. This higher carbon dioxide concentration has a stimulating effect on the neurons of the respiratory center in the medulla. They send out more impulses to the respiratory muscles, and as a result, respirations increase, becoming both faster and deeper. Therefore, as more air moves in and out of the lungs per minute, more carbon dioxide leaves the blood, more oxygen enters it, and more energy becomes available for doing the extra work of exercise.

Another important automatic adjustment that helps supply cells with more oxygen when they are doing more work has to do with the circulatory system instead of the respiratory system. The heart pumps more blood through the body per minute because it beats faster and harder. This means that the millions of red blood cells make more round trips between the lungs and tissues each minute and so deliver more oxygen per minute to tissue cells.

DISORDERS OF THE RESPIRATORY SYSTEM

Many different kinds of diseases and injuries may be responsible for respiratory disorders. For example tuberculosis may destroy part of the lungs; lung cancer may do the same thing; and pneumonia may plug up alveoli. A brain hemorrhage may depress the respiratory center, causing respirations to become slow and labored or even to stop completely. In recent years, a disease rarely heard of in our grandfathers' generation—*emphysema*—has become increasingly common. In this disease the walls of many of the alveoli become greatly overstretched. They do not collapse during expiration as normal alveoli do, so that air remains trapped in them instead of being exhaled. As a result, less air than normal is exhaled and inhaled and less oxygen and carbon dioxide are exchanged between alveolar air and blood. Emphysema victims, therefore, become *hypoxic*, that is, their cells receive less oxygen than what they need for normal functioning. Lung disease is not the only abnormality that can produce hypoxia. Anemia, for example, can also cause hypoxia. An anemic individual's red blood cells contain less hemoglobin than normal and so can transport less oxygen than normal.

Outline summary—The respiratory system

Basic plan similar to an inverted tree if it were hollow; leaves of tree would be comparable to alveoli, the microscopic sacs enclosed by networks of capillaries

NOSE
1. Structure
   a. Nasal septum separates interior of nose into two cavities
   b. Mucous membrane lines nose
   c. Frontal, maxillary, sphenoid, and ethmoid sinuses drain into nose
2. Functions
   a. Warms and moistens air inhaled
   b. Contains sense organs of smell

PHARYNX
1. Structure
   a. Two nasal cavities, mouth, esophagus, larynx, and eustachian tubes all have openings into pharynx
   b. Tonsils and adenoids located in pharynx
   c. Mucous membrane lines pharynx
2. Functions
   a. Passageway for food and liquids
   b. Passageway for air

LARYNX (VOICE BOX)
1. Structure
   a. Several pieces of cartilage form framework
   b. Mucous lining
   c. Vocal cords stretch across interior of larynx
2. Functions
   a. Passageway for air to and from lungs
   b. Voice production

TRACHEA (WINDPIPE)
1. Structure
   a. Mucous lining
   b. C-shaped rings of cartilage hold trachea open
2. Function—passageway for air to and from lungs

THE BRONCHI, BRONCHIOLES, AND ALVEOLI
1. Structure
   a. Trachea branches into right and left bronchi
   b. Each bronchus branches into smaller and smaller tubes called bronchioles
   c. Bronchioles end in clusters of microscopic alveolar sacs, the walls of which are made up of alveoli
2. Function
   a. Bronchi and bronchioles—passageway for air to and from alveoli
   b. Alveoli—exchange of gases between air and blood

THE LUNGS
1. Structure
   a. Size—large enough to fill chest cavity except for middlespace occupied by heart and large blood vessels
   b. Apex—narrow upper part of each lung, under collarbone
   c. Base—broad lower part of each lung; rests on diaphragm
   d. Pleura—moist smooth slippery membrane that lines chest cavity and covers outer surface of lungs; prevents friction between lungs and chest wall during breathing
2. Function—breathing (respiration)

RESPIRATION
Exchange of gases in lungs
1. Carbohemoglobin breaks down into carbon dioxide and hemoglobin
2. Carbon dioxide moves out of lung capillary blood into alveolar air and out of body in the expired air
3. Hemoglobin combines with oxygen, producing oxyhemoglobin

Exchange of gases in tissues
1. Oxyhemoglobin breaks down into oxygen and hemoglobin

2. Oxygen moves out of tissue capillary blood into tissue cells
3. Carbon dioxide moves from tissue cells into tissue capillary blood
4. Hemoglobin combines with carbon dioxide, forming carbohemoglobin

Mechanics of breathing

1. Contraction of diaphragm and of chest-elevating muscles enlarges chest cavity, expands lungs, and causes air to move down into lungs
2. Relaxation of diaphragm and of chest elevators decreases size of chest cavity, deflates lungs, and causes air to move out of lungs
3. Tidal air—amount normally breathed in and out with each breath
4. Vital capacity—largest amount of air we can breathe in and out with one breath
5. Rate—usually about 16 to 20 breaths a minute; much faster during exercise

DISORDERS OF THE RESPIRATORY SYSTEM

1. Tuberculosis
2. Lung cancer
3. Pneumonia
4. Brain hemorrhage
5. Emphysema

Review questions—The respiratory system

1. What and where are the pharynx and larynx?
2. What and where are alveoli and the pleura?
3. Why does sinusitis frequently follow a common cold?
4. What and where are the eustachian tubes?
5. Where are the tonsils and adenoids located?
6. What and where is the "Adam's apple"?
7. What does the term "vital capacity" mean?

VIII

The urinary system

The urinary system, as you might guess from its name, performs the functions of secreting urine and eliminating it from the body. What you might not guess so easily is how essential these functions are for healthy survival. Unless they operate normally, the normal composition of blood cannot long be maintained, and serious consequences soon follow. In this chapter we shall discuss the structure and function of each of the urinary system's organs. We shall also mention briefly some disease conditions produced by abnormal functioning of the urinary system.

THE KIDNEYS
Location and structure

There are two kidneys. They lie behind the abdominal organs against the muscles of the back. Usually the left kidney is a little larger than the right and a little farther above the waistline. A heavy cushion of fat normally encases each kidney and helps hold it up in place; so in a very thin person, the kidneys may drop down a little (renal ptosis). This hinders urine drainage by putting a kink in the ureter, the tube that drains urine out of the kidneys.

The outer part of the kidney is called the *cortex*. (The word "cortex" comes from the Latin word that means bark or rind, so the cortex of an organ—of the kidney, the brain, the adrenal glands—is its outer layer.) The interior part of the kidney, beneath the cortex, is the *medulla*.

Kidney cells and capillaries are arranged so that they form unique little structures called *nephrons*. Each nephron consists of three main parts: a glomerulus, a Bowman's capsule, and a tubule. The *glomerulus* is a network of blood capillaries tucked into the top of a microscopic-sized funnel-shaped structure. The top part of this structure is called the *Bowman's capsule*, and

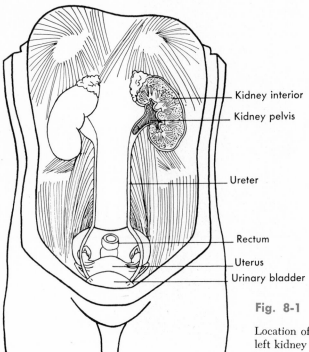

Kidney interior
Kidney pelvis

Ureter

Rectum
Uterus
Urinary bladder

Fig. 8-1

Location of the urinary organs. The surface of the left kidney and the upper part of its ureter are cut away to show the interior.

the stem part is called the *renal tubule.* As you can see in Figure 8-2, different parts of the renal tubule have different names. The first segment of each tubule is called the proximal convoluted tubule—proximal because it lies nearest the tubule's origin from the Bowman's capsule, and convoluted because it twists around to form several coils. The proximal tubule becomes the descending limb and then the ascending limb of the loop of Henle. The ascending limb becomes the distal convoluted tubule, which terminates in a straight or collecting tubule that opens into the renal pelvis.

Functions

The structure of nephrons makes them able to carry on their function of urine formation. Because the walls of the glomerular capillaries and of the Bowman's capsules are very thin membranes, water and dissolved substances (except albumin and other blood proteins) filter rapidly out of the blood in the glomeruli into the Bowman's capsules. This filtrate then trickles down the convoluted tubules, and as it does so a large part of the water goes back into the blood, that is, is reabsorbed into capillaries around the tubules. Dissolved substances (solutes) also leave the tubule filtrate to return to the blood. Glucose, for instance, is entirely reabsorbed, so that none of it is wasted by being lost in the urine. Exceptions to this normal rule, however, do occur. For example, in diabetes mellitus, if blood glucose concentration increases above a certain level, the tubular filtrate then contains more glucose than the kidney tubule cells can reabsorb into the blood. Some of the glucose therefore remains behind in the urine; sugar in the urine (glycosuria or glucosuria) is a common sign of diabetes.

Like glucose, sodium chloride and other salts filter out of glomerular blood. Unlike glucose, however, salts are only partially reabsorbed from the tubule filtrate. The

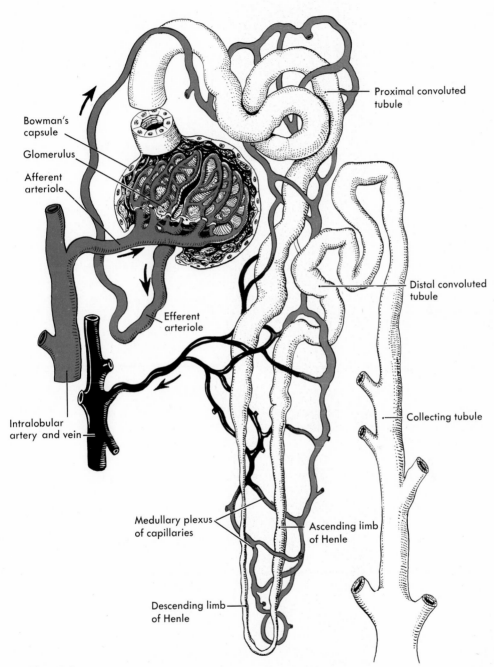

Fig. 8-2

Blood vessels of the nephron unit. Arrows indicate the direction of blood flow: from the intra-lobular artery ⟶ the afferent arteriole ⟶ the glomerulus ⟶ the efferent arteriole ⟶ the peritubular capillaries (around the tubules) ⟶ the venules (shown in black) ⟶ the intralobular vein.

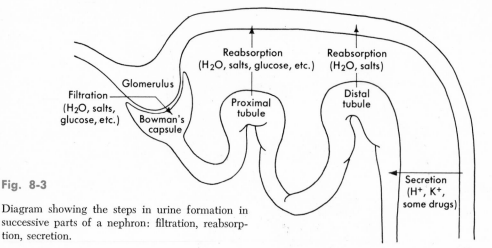

Fig. 8-3

Diagram showing the steps in urine formation in successive parts of a nephron: filtration, reabsorption, secretion.

amount reabsorbed varies from time to time. When, for example, you salt food heavily, your kidneys reabsorb less salt than when you use salt sparingly. Thus your body gets rid of excess salt by way of the urine, and in this way tends to keep the blood's salt concentration normal. This is an extremely important matter because cells are damaged by either too much or too little salt in the fluids around them.

By way of summary, three processes in succession accomplish the task of urine formation (see Figure 8-3):

1. *Filtration*—of water and dissolved substances out of the blood in the glomeruli into the Bowman's capsules;
2. *Reabsorption*—of water and dissolved substances out of the kidney tubules back into the blood (Note that this process prevents substances needed by the body from being lost in the urine. Usually 97% to 99% of the water filtered out of the glomerular blood is retrieved from the tubules.);
3. *Secretion*—of hydrogen ions (H^+), potassium ions (K^+), and certain drugs (for example, penicillin).

Secretion, like reabsorption, is also carried on by kidney tubules. Substances secreted move out of blood into tubule urine—a kind of reabsorption in reverse since sub-

stances reabsorbed move out of tubule urine into blood. Kidney tubule secretion is an essential process. Kidney tubules must secrete varying amounts of the acid hydrogen ions in order to maintain the slightly alkaline reaction of blood that is essential for healthy survival.

The body has ways to control both the amount and the composition of the urine that it secretes. It does this mainly by controlling the amount of water and dissolved substances reabsorbed by the convoluted tubules. A hormone (ADH or antidiuretic hormone) from the posterior pituitary gland, for instance, tends to decrease the amount of urine by causing tubule cells to reabsorb more water. As a result, less water is lost from the body or more water is retained—whichever way you wish to say it. At any rate, for this reason ADH is accurately described as the "water-retaining hormone." The hormone aldosterone secreted by the adrenal cortex plays an important part in controlling the kidney tubules' reabsorption of salt. Because aldosterone primarily stimulates the tubules to reabsorb sodium salts at a faster rate, it is sometimes called the "salt-retaining hormone." Secondarily aldosterone tends also to increase tubular water reabsorption.

Sometimes the kidneys do not excrete

normal amounts of urine; for example, as a result of kidney disease, cardiovascular disease, certain hormonal imbalances, or other factors. (See question 6, page 114.) Here are some terms associated with abnormal amounts of urine:

1. *Anuria*—literally, absence of urine;
2. *Oliguria*—scanty urine;
3. *Polyuria*—unusually large amounts of urine.

THE URETERS

Urine drains out of the collecting tubules of each kidney into the renal pelvis and on down the ureter into the urinary bladder (Figure 8-1). The *renal pelvis* is the basin-like upper end of the ureter and is located inside the kidney. Ureters are narrow tubes less than ¼ inch wide but 10 to 12 inches long. Mucous membrane lines both ureters and renal pelves. Smooth muscle fibers in the ureter walls contract to produce a peristalic movement that forces urine down into the bladder.

THE URINARY BLADDER

Elastic fibers and involuntary muscle fibers in the wall of the urinary bladder make it well suited for its functions of expanding to hold variable amounts of urine and then contracting to empty itself. Most people feel the desire to void when the bladder contains about ½ pint (250 milliliters) of urine. Mucous membrane lines the urinary bladder. It lies in folds (rugae) when the bladder is empty.

Infection of the mucous membrane lining the urinary tract is not uncommon. Have you ever heard of cystitis or pyelitis? *Cystitis* is an infection of the bladder lining; *pyelitis* is an infection of the lining of the renal pelvis.

Urinary *retention* is a condition in which no urine is voided. The kidneys secrete urine but the bladder for one reason or another cannot empty itself. In urinary *suppression*, the opposite is true. The kidneys do not secrete any urine, but the bladder retains the ability to empty itself. Do you think anuria could be a symptom of either suppression or retention? Why?

Incontinence is a condition in which the patient voids urine involuntarily. It frequently occurs in patients who have suffered a stroke or spinal cord injury.

THE URETHRA

To leave the body, urine passes from the bladder down the urethra and out its external opening, the *urinary meatus*. In other words, the urethra is the lowermost part of the urinary tract. The same sheet of mucous membrane that lines the renal pelves, ureters, and bladder extends down into the urethra too—an interesting structural feature because it accounts for the fact that an infection of the urethra may spread upward throughout the urinary tract. The urethra is a narrow tube. It is only about 1½ inches long in a woman but about 8 inches long in a man.

Outline summary—The urinary system

KIDNEYS
Structure
1. Cortex—outer layer of kidney
2. Medulla—interior part of kidney
3. Nephrons—microscopic functional units of kidneys; each consists of a glomerulus, a Bowman's capsule, and a renal tubule

Function
1. Form urine by means of glomerular filtration, tubule reabsorption, and tubule secretion
2. Amount of urine controlled mainly by hormones—ADH—from posterior pituitary gland and aldosterone from adrenal cortex

URETERS
1. Structure
   a. Narrow long tubes with expanded up-

per end (renal pelvis) located inside kidney
   b. Mucous lining
2. Function—drain urine from kidneys to urinary bladder

URINARY BLADDER
1. Structure
   a. Elastic muscular organ, capable of great expansion
   b. Lined with mucous membrane arranged in rugae, like stomach mucosa
2. Functions
   a. Store urine before voiding
   b. Voiding

THE URETHRA
1. Structure
   a. Narrow short tube from urinary bladder to exterior
   b. Lined with mucous membrane
   c. Opening of urethra to exterior called urinary meatus
2. Function
   a. Serves as passageway by which urine leaves bladder for exterior
   b. Passageway from which male reproductive fluid leaves body

Review questions—The urinary system

1. What organs form the urinary system?
2. What and where are the glomeruli and Bowman's capsules?
3. Explain briefly the functions of the glomeruli and Bowman's capsules.
4. Explain briefly the function of the renal tubules.
5. Explain briefly the function of ADH. What is the full name of this hormone? What gland secretes it?
6. Suppose that ADH secretion increases markedly. Would this increase or decrease urine volume? Why?
7. What kind of membrane lines the urinary tract?
8. What is the urinary meatus?
9. What and where are the ureters and the urethra?
10. Explain briefly the meaning of each of the following terms:

anuria	polyuria
cystitis	pyelitis
incontinence	urinary retention
nephritis	urinary suppression
oliguria	

How the body provides for survival of the human species

IX

The reproductive systems

"Fearfully and wonderfully made" we truly are. Almost any one of the body's structures or functions might have inspired this statement, but of them all perhaps the reproductive systems best deserve such praise. Their achievement? The miracle of duplicating the human body. Their goal? The survival of the human species.

As you probably noticed, we used the plural, reproductive systems, in the preceding paragraph and the chapter title. The male reproductive system consists of one group of organs and the female reproductive system of another group. These two systems differ in structure, but they share a common function—that of reproducing the human body. First the male reproductive system will be discussed and then the female reproductive system.

THE MALE REPRODUCTIVE SYSTEM
Structural plan

So many organs make up the male reproductive system that we need to look first at the structural plan of the system as a whole. It consists of a pair of main sex glands, a series of ducts from these to the exterior, accessory sex glands, and external reproductive organs. Table 9-1 lists the names of all of these structures and Figure 9-1 shows the location of most of them.

The testes

To judge the testes by their size would be to underestimate their importance. Each of these two oval-shaped glands is only about 1½ inches long and 1 inch wide, yet each testis forms millions of male sex cells (*spermatozoa*, or *sperm*), any one of which

117

Table 9-1. Male reproductive organs

Main sex gland	Testes (right testis and left testis)
Ducts	Epididymis, seminal duct, ejaculatory duct (two of each of the preceding), and one urethra
Accessory sex glands	Seminal vesicle, bulbourethral gland (two of each of the preceding), and one prostate gland
External genitals	Scrotum and penis

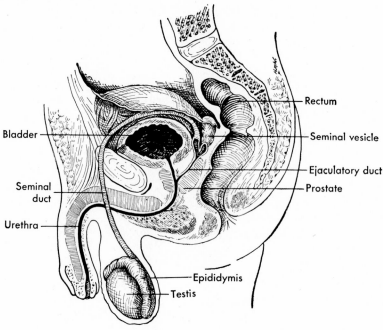

Fig. 9-1

Longitudinal section of the male pelvis, showing the location of the male reproductive organs.

may join with a female sex cell (ovum) to become a new human being. Each testis also secretes the male sex hormone (testosterone), the hormone that in a few short months transforms a little boy into a man. Testosterone lowers the pitch of his voice; it makes his muscles grow large and strong; it even changes the size and shape of his bones.

The testes are located in an external organ—a pouchlike, skin-covered structure called the *scrotum*. Each testis consists of several sections (lobules), and each lobule consists of a narrow but long and coiled *seminiferous tubule*. From the age of puberty on, the seminiferous tubules are almost continuously forming spermatozoa or sperm. *Spermatogenesis* is the name of this process. The other function of the testes is to secrete the male hormone, testosterone. This function, however, is carried on by the *interstitial cells* of the testes, not by its seminiferous tubules. Numerous clusters of interstitial cells are located in the connective tissue between the seminiferous tubules.

The epididymis, seminal duct, ejaculatory duct, and urethra

From the seminiferous tubules, sperm by the millions stream into a narrow but long and tightly coiled duct, the *epididymis*. An epididymis lies over the top of each testis. By examining Figure 9-1 you can trace the course taken by the sperm in reaching the exterior. From the epididymis, sperm move on into a seminal duct and then into an *ejaculatory duct,* and finally into the *urethra*. Note that in the male, the urethra serves as the outlet for both the urinary tract and the reproductive tract. This is not the case in the female. In a woman's body, the urethra serves only the urinary tract and a separate tube, the vagina, serves the reproductive tract.

The seminal vesicles, prostate gland, and bulbourethral glands

The two seminal vesicles, one prostate gland, and two bulbourethral glands are accessory male glands that produce alkaline secretions. These secretions plus the sperm constitute the *seminal fluid* or *semen.*

The *prostate gland* claims importance not so much for its function as for its troublemaking. In older men it often becomes inflamed and enlarged, squeezing on the urethra, which runs through the center of the doughnut-shaped prostate. Sometimes, in fact, the prostate enlarges so much that it closes off the urethra completely. Urination then becomes impossible. (Should you refer to this as urinary retention or urinary suppression? If you are not sure, check your answer on page 113.)

The small *bulbourethral* (Cowper's) *glands* lie one on either side of the urethra just below the prostate gland. Like the seminal vesicles and the prostate, the bulbourethral glands add an alkaline secretion to the semen. Sperm survive and remain fertile longer in an alkaline fluid than in an acid one.

The scrotum and penis

Male external reproductive organs *(genitals)* consist of two organs, the scrotum and the penis. A special kind of tissue known as erectile tissue composes most of the interior of the penis. Erectile tissue contains many small spaces that usually are collapsed. Under the stimulus of the sexual emotion, blood floods these spaces, distending them enough to cause enlargement and rigidity of the organ.

Skin covers over the outside of the penis. At the lower end of this organ, the skin is folded over double to form a somewhat loose-fitting casing called the foreskin (or prepuce). Circumcision involves cutting the foreskin so that it will not fit too tightly and cause irritation.

THE FEMALE REPRODUCTIVE SYSTEM
Structural plan

The structural plan of the female reproductive system resembles that of the male

Table 9-2. The female reproductive system

Main sex gland	Ovaries
Ducts	Uterine tubes, uterus, and vagina
Accessory glands	Bartholin's glands
External genitals	Vulva (or pudendum)

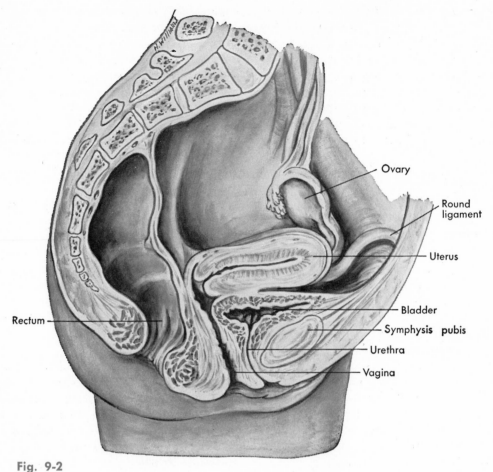

Fig. 9-2

Longitudinal section of the female pelvis, showing the location of the female reproductive organs.

system. Like the male reproductive system, the female reproductive system consists of a pair of main sex glands, ducts from these to the exterior, accessory sex glands, and external genitals. To find out the names of these structures in the female, consult Table 9-2.

The ovaries

Although male and female reproductive systems resemble each other in plan, they differ from one another in details. For example, the testes, the main sex glands of the male, are not located inside a body cavity but lie in an external skin-covered pouch, the scrotum. In the female, on the other hand, the main sex glands, the ovaries, lie within the pelvic cavity. There is also another difference from the male—no duct attaches directly to the ovary as one does to the testis. More will be said about the importance of this fact later (page 121).

Several thousand sacs, too small to be seen without a microscope, make up the bulk of each ovary. They are called *graafian follicles*, in honor of a Dutch anatomist who lived some 300 years ago. Within each follicle lies an immature *ovum*, the female sex cell. One ovum usually matures each month during the 30 or 40 years that a woman has menstrual periods. As the ovum is maturing, the cells of the follicle secrete one kind of female hormone—estrogens. The follicle gradually enlarges, and as it does so it moves to the surface of the ovary. Here it breaks open, ejecting the ovum into the pelvic cavity, not into a duct. This process is called *ovulation*. When does ovulation occur? Probably most physiologists today would answer that it usually occurs fourteen days before the next menstrual period begins. But they would quickly add that there are exceptions to this general rule. Sometimes ovulation takes place much earlier, presumably as much as nineteen days or more before the beginning of the next menses. This matter of the time of ovulation has great practical importance because of its relation to birth control. An ovum lives only a short time after it is expelled from its follicle, and a sperm lives only a short time after it enters the female body. Fertilization of the ovum by the sperm, therefore, can occur only around the time of ovulation. In other words, a woman's fertile period lasts only a few days out of each month. This knowledge forms the basis of the rhythm method of birth control.

After ovulation, the torn follicle turns a golden color; it is then called the *corpus luteum*. Cells of the corpus luteum, like cells of the graafian follicles, are endocrine glands. They secrete hormones, but unlike the follicles, which secrete only one kind of hormone (estrogens), the corpus luteum secretes two kinds of female hormones— progesterone and estrogens. What eventually happens to the corpus luteum depends upon whether or not a sperm unites with the ovum within the first 24 hours or so after ovulation. If this has not occurred, the corpus luteum stops secreting progesterone about twelve days after ovulation and also slows down on the secretion of estrogens. It shrinks in size and soon disappears, leaving only a small scar on the surface of the ovary. If fertilization has taken place, the corpus luteum continues to secrete both of its hormones during the pregnancy. These are essential for the maintenance of pregnancy, especially during the early months.

In reading the preceding two paragraphs did you notice two functions that the ovaries perform? Like the testes, they produce both sex cells and sex hormones. However, the ovaries usually produce only one mature sex cell a month, whereas the testes produce hundreds of millions. The ovaries produce two kinds of female hormones— estrogens and progesterones. The testes, on the other hand, produce only one important male hormone—testosterone. We shall save our discussion of the functions of male and female hormones for the next chapter.

The uterine tubes (fallopian tubes)

The uterine tubes serves as ducts for the ovaries even though they are not attached to them. The outer end of each tube curves over the top of each ovary. The inner end of each uterine tube attaches to the uterus, and the cavity inside the tube opens into the cavity in the uterus. Each tube measures about 4 inches in length.

After ovulation, the ovum finds its way into one of the uterine tubes. This is where fertilization (union of a sperm with an ovum) normally occurs. Occasionally, however, because the tubes are not actually connected to the ovaries, an ovum does not enter a tube but becomes fertilized in the pelvic cavity. The term *ectopic pregnancy* means a pregnancy that develops outside of its proper place in the cavity of the uterus.

The uterus

The uterus is a small organ—only about the size of a pear—but it is extremely strong. It is almost all muscle with only a small cavity inside. During pregnancy the uterus grows many times larger so that it

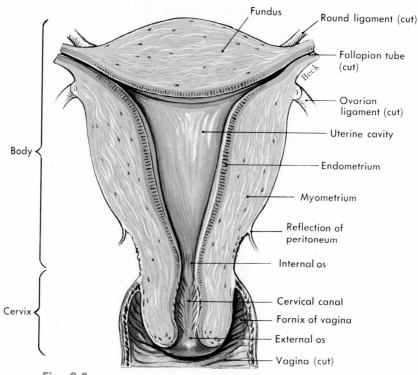

Fundus
Round ligament (cut)
Fallopian tube (cut)
Ovarian ligament (cut)
Uterine cavity
Endometrium
Myometrium
Reflection of peritoneum
Internal os
Cervical canal
Fornix of vagina
External os
Vagina (cut)
Body
Cervix
Beck

Fig. 9-3

Sectioned view of the uterus.

becomes big enough to hold a baby plus considerable fluid. The uterus is composed of two parts: an upper portion, the *body,* and a lower narrow section, the *cervix.* Just above the level where the uterine tubes attach to the body of the uterus, it rounds out to form a bulging prominence called the *fundus.* Except during pregnancy, the uterus lies in the pelvic cavity just behind the urinary bladder. By the end of pregnancy it becomes large enough to extend up to the top of the abdominal cavity. It then presses against the underside of the diaphragm—a fact that explains such a comment as "I can't seem to take a deep breath since I've gotten so big," made by many women late in their pregnancies.

The uterus functions in three processes—menstruation, pregnancy, and labor. As already noted, the corpus luteum stops se-

creting progesterone and slows down on the secretion of estrogens about twelve days after ovulation. Two days later, when the blood progesterone and estrogen concentration are at their lowest, menstruation starts. Bits of *endometrium* (mucous membrane lining of the uterus) pull loose, leaving torn blood vessels underneath. Blood and bits of endometrium trickle out of the uterus into the vagina and out of the body. Immediately after menstruation the endometrium starts to repair itself. It again grows thick and becomes lavishly supplied with blood in preparation for pregnancy. But if fertilization does not take place, the uterus once more sheds the lining made ready for a pregnancy that did not occur. Because these changes in the uterine lining continue to repeat themselves over and over, they are spoken of as the *menstrual cycle.* For a description of this cycle in the form of a diagram, see Figure 9-4.

Menstruation first occurs at puberty, often

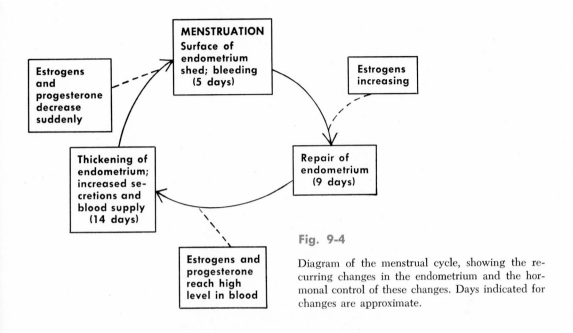

MENSTRUATION
Surface of
endometrium
shed; bleeding
(5 days)

Estrogens
and
progesterone
decrease
suddenly

Estrogens
increasing

Thickening of
endometrium;
increased se-
cretions and
blood supply
(14 days)

Repair of
endometrium
(9 days)

Estrogens and
progesterone
reach high
level in blood

Fig. 9-4

Diagram of the menstrual cycle, showing the re-
curring changes in the endometrium and the hormonal control of these changes. Days indicated for
changes are approximate.

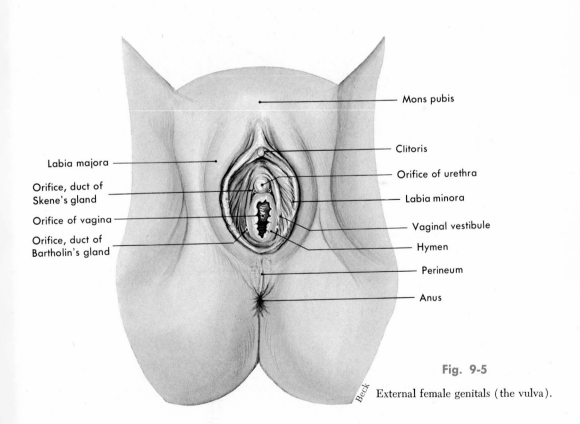

Mons pubis

Clitoris

Orifice of urethra

Labia minora

Vaginal vestibule

Hymen

Perineum

Anus

Labia majora

Orifice, duct of
Skene's gland

Orifice of vagina

Orifice, duct of
Bartholin's gland

Fig. 9-5

External female genitals (the vulva).

around the age of 13 years. Normally, it repeats itself about every 28 days or thirteen times a year for some 30 years or so before it ceases (menopause or climacteric), when a woman is somewhere around the age of 45 years.

The vagina

The vagina is a distensible tube made mainly of smooth muscle and lined with mucous membrane. It lies in the pelvic cavity between the urinary bladder and the rectum. As the part of the female reproduc-

tive tract that opens to the exterior, the vagina is the organ that sperm enter on their journey to meet an ovum, and it is also the organ from which a baby emerges to meet its new world.

Bartholin's glands

One of the small Bartholin's glands lies to the right of the vaginal outlet and one to the left of it. Their ducts open into the space between the labia minora and the hymen. Bartholinitis, an infection of these glands, occurs frequently. It often develops,

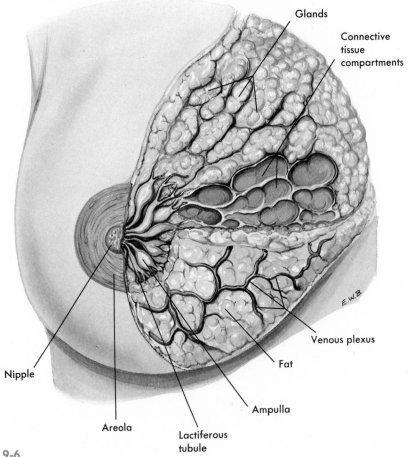

Fig. 9-6

Anterior view of the breast. The skin has been removed in the lower right quadrant to reveal the venous plexus overlying the adipose (fatty) tissue. In the upper right quadrant the adipose tissue has been removed to show the alveoli of the glands. The glands have been removed in a small area to reveal the connective tissue compartments that separate the lobules.

for example, when a woman contracts gonorrhea.

The vulva

See Figure 9-5.

The breasts

Each breast consists of fifteen to twenty-five divisions or lobes that are arranged radially. Each lobe consists of several lobules, and each lobule consists of secreting cells. Grapelike clusters of the secreting cells surround small ducts. The small ducts unite so that only one duct leads from each lobe to an opening in the nipple. The colored area around the nipple is the *areola*. It changes from a delicate pink to brown early in pregnancy and never quite returns to its original color.

DISORDERS OF THE REPRODUCTIVE SYSTEMS

Infections of the mucous membrane lining of the reproductive tract occur in both sexes. In women the danger of this becomes especially great following childbirth. Physicians and nurses therefore take great care not to introduce infectious organisms into the reproductive tract during or after delivery. The name given the inflammation indicates the part of the tract inflamed. For example, vaginitis is inflammation of the vagina; cervicitis, inflammation of the cervix; endometritis, inflammation of the endometrium; salpingitis, inflammation of the uterine tubes. Tumors frequently develop in the uterus, ovaries, breasts, and prostate gland.

Outline summary—The reproductive systems

MALE REPRODUCTIVE SYSTEM
Structural plan

Consists of pair of main sex glands, series of ducts, accessory sex glands, and external reproductive organs (genitals); see Table 9-1, page 118

Structures

1. Testes—pair of small oval glands in scrotum; main male sex glands; seminiferous tubules form male sex cells (spermatogenesis); interstitial cells secrete male hormone (testosterone)
2. Epididymis—narrow tube attached to each testis; duct of testis
3. Seminal duct—continuation of duct that starts with epididymis
4. Ejaculatory duct—continuation of seminal duct
5. Urethra—terminal duct of male reproductive tract as well as of urinary tract
6. Prostate gland—encircles urethra just below bladder; secretes alkaline fluid, part of semen
7. Bulbourethral glands—pair of small glands located just below prostate; ducts open into urethra where they add alkaline secretion to semen
8. Scrotum and penis—external genitals

FEMALE REPRODUCTIVE SYSTEM
Structural plan
Same as male; see Table 9-2, page 119

Structures

1. Ovaries—main sex glands of female; located in pelvic cavity; female sex cells (ova) form in graafian follicles in ovaries; graafian follicles secrete estrogens and corpus luteum secretes estrogens and progesterone
2. Uterine tubes (fallopian tubes)—ducts for ovaries but not attached to them
3. Uterus (womb)
   a. Parts—fundus, body, cervix; see Figure 9-3
   b. Structure—strong muscular organ with mucous lining (endometrium)
   c. Functions—menstruation, pregnancy, labor; see Figure 9-4 for diagram of menstrual cycle
4. Vagina—muscular tube lined with mucous membrane; terminal part of female reproductive tract
5. Bartholin's glands—pair of small glands near vaginal orifice

6. Vulva—female external genitals
7. Breasts—glands present in both sexes but normally function only in female

DISORDERS OF THE REPRODUCTIVE SYSTEMS

1. Infections—common in both sexes; for example, vaginitis, endometritis, prostatitis
2. Tumors—occur in both sexes; for example, in uterus, ovaries, breasts, and prostate

Review questions—The reproductive systems

1. What and where is each of the following:

   prostate gland   cervix
   ovary   ova
   spermatozoa   Bartholin's glands
   testis   graafian follicles
   epididymis

2. What is ovulation?
3. When does ovulation occur with relation to menstruation?
4. Name the female hormones and the glands that secrete them.
5. Name the male hormone. What glands secrete it?
6. What is the corpus luteum? What is its function?
7. What is an ectopic pregnancy? What structural fact about the female reproductive system makes an ectopic pregnancy possible?
8. Explain what menstruation is. What hormones control the time of menstruation?
9. What does each of the following terms mean? cervicitis, endometritis, and salpingitis.
10. How many female sex cells are usually formed each month? How does this compare with the number of male sex cells formed each month?

How the endocrine system controls body functions

X

The endocrine system

Have you ever seen a giant or a dwarf? Have you ever known anyone who had sugar diabetes or a goiter? If so, you have had visible proof of the importance of the endocrine system for normal development and health.

The organs of the endocrine system are located in widely separated parts of the body—in the cranial cavity, in the neck, in the thoracic cavity, in the abdominal cavity, in the pelvic cavity, and even outside any body cavity. Note the names and locations of the endocrine glands shown in Figure 10-1.

All the organs of the endocrine system are glands, but all glands are not organs of the endocrine system. Those glands that discharge their secretions into ducts are not endocrine glands. Endocrine glands are those which secrete chemicals known as hormones into the blood.

The endocrine system performs the same general functions as the nervous system—namely, communication and control. As you know, the nervous system sends nerve impulses by way of nerve fibers to organs in order to regulate their activities. The endocrine system, in contrast, sends hormones by way of the circulating blood.

In this chapter you will read about the main functions of individual endocrine glands and will learn of the importance of the endocrine system for healthy survival.

THE PITUITARY BODY (OR HYPOPHYSIS)

The pituitary body, an organ no larger than a pea, is really two endocrine glands.

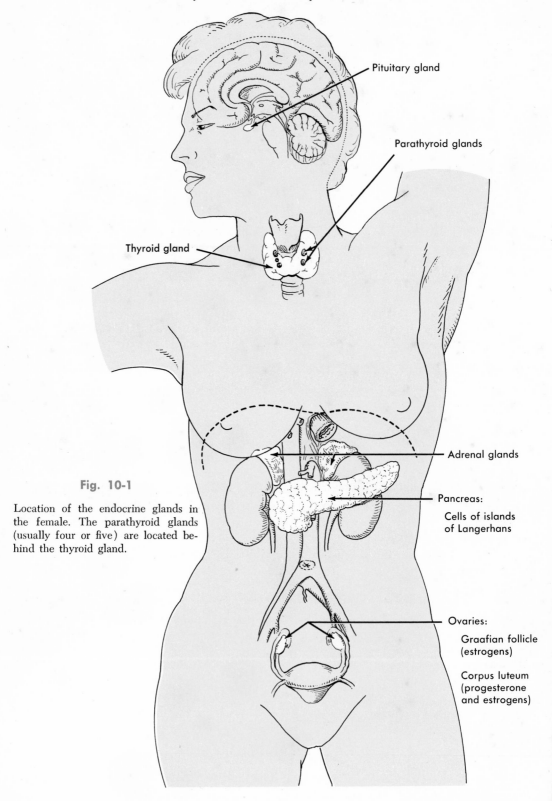

Pituitary gland

Parathyroid glands

Thyroid gland

Adrenal glands

Pancreas:

Cells of islands
of Langerhans

Ovaries:

Graafian follicle
(estrogens)

Corpus luteum
(progesterone
and estrogens)

Fig. 10-1

Location of the endocrine glands in
the female. The parathyroid glands
(usually four or five) are located behind the thyroid gland.

One is called the anterior pituitary gland (or adenohypophysis) and the other is called the posterior pituitary gland (or neurohypophysis). The location of the pituitary body, because it is such a protected one, suggests this organ's importance. The pituitary body lies buried deep in the cranial cavity, in the small depression of the sphenoid bone which is shaped like a saddle and is called Turk's saddle or the sella turcica. A stemlike structure, the pituitary stalk, attaches the gland to the undersurface of the brain. More specifically, the stalk attaches the pituitary body to the hypothalamus.

Functions of anterior pituitary gland hormones

The anterior pituitary gland secretes seven hormones. Six of them are usually referred to by their abbreviated names— TH, ACTH, FSH, LH, GH, and MSH. Learning these is a little like learning a mixed-up alphabet, but if you associate the letters with the full names of the hormones (given in the next paragraphs), you probably will find the task fairly easy.

The anterior pituitary gland is often called the master gland—because four anterior pituitary hormones act on four other endocrine glands to control both their structure and functioning. The four hormones which act in this way are: thyrotrophic hormone, adrenocorticotrophic hormone, follicle-stimulating hormone, and luteinizing hormone.

Thyrotrophic hormone (TH) acts on the thyroid gland. It promotes and maintains its normal growth and development, and it stimulates the thyroid to secrete thyroid hormone, which is made up of thyroxin and triiodothyronine.

Adrenocorticotrophic hormone (ACTH) acts on the adrenal cortex. It promotes and maintains its normal growth and development and stimulates it to secrete glucocorticoids—mainly the hormone called cortisol or hydrocortisone.

Follicle-stimulating hormone (FSH) stimulates primary graafian follicles to start growing and to continue developing to maturity, that is, to the point of ovulation. FSH also stimulates follicle cells to secrete estrogens. In the male, FSH promotes and maintains normal growth and development of the seminiferous tubules; it also maintains spermatogenesis by them.

Luteinizing hormone (LH) performs four functions. It acts with FSH to stimulate a follicle and ovum to complete their growth to maturity. It stimulates follicle cells to secrete estrogens. It is essential for ovulation (the rupturing of the mature follicle with expulsion of its ripe ovum.) Because of this function, LH is sometimes called the *ovulating hormone*. Finally, LH stimulates the formation of a golden body, the corpus luteum, in the ruptured follicle— the process called *luteinization*. This function, obviously, is the one which earned LH its title of luteinizing hormone.

The other three hormones secreted by the anterior pituitary gland are growth hormone, lactogenic hormone, and melanocyte-stimulating hormone. *Growth hormone* (GH), also known as STH or somatotrophic hormone, tends to speed up the movement of digested proteins (amino acids) out of the blood and into the cells. This tends to accelerate the cells' anabolism of amino acids to form tissue proteins; hence this action promotes normal growth. Growth hormone also affects fat and carbohydrate metabolism: it accelerates fat catabolism but slows glucose catabolism. This means that less glucose leaves the blood to enter cells and that, therefore, the amount of glucose in the blood tends to increase. Thus growth hormone and insulin tend to have opposite effects on blood sugar. Insulin tends to decrease blood sugar, and growth hormone tends to increase it. An interesting application of this fact occurs in the disease called acromegaly. This is an adult disease in which excessive amounts of growth hormone are secreted and in which the blood glucose concentration is usually abnormally

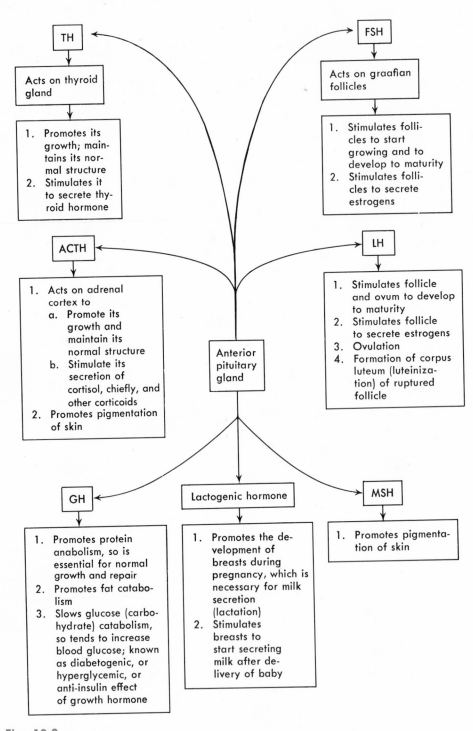

Fig. 10-2

Diagram of the functions of the anterior pituitary hormones.

high (hyperglycemia). This type of hyper-glycemia is appropriately referred to as pituitary diabetes. Growth hormone is also referred to as a diabetogenic or hypergly-cemic hormone because of its effects on carbohydrate metabolism.

Lactogenic hormone or prolactin, as its name suggests, generates or initiates milk secretion. During pregnancy, lactogenic hormone stimulates the breast development necessary for eventual milk secretion, or lactation. Soon after the delivery of a baby, lactogenic hormone stimulates the breasts to start secreting milk.

Melanocyte-stimulating hormone (MSH) tends to increase pigmentation of the skin. Melanocytes are cells that form melanin, the pigment of the skin. Recently scientists have established that another anterior pitui-tary hormone, ACTH, also tends to increase skin pigmentation.

For a brief summary of anterior pitui-tary hormones and their functions, see Fig-ure 10-2, page 132.

Functions of posterior pituitary gland hormones

The posterior pituitary gland secretes two hormones—antidiuretic hormone (ADH) and oxytocin. ADH stimulates kidney tu-bules to reabsorb water back into the blood faster. With more water moving out of the tubules into the blood, less water of course remains in the tubules and therefore less urine leaves the body. Incidentally, this is the reason the name antidiuretic hormone is an appropriate one—because "anti" means "against" and "diuretic" means "increasing the secretion of urine."

The posterior pituitary hormone, oxyto-cin, is only secreted by a woman's body before and after she has a baby. Oxytocin stimulates contraction of the smooth mus-cle of the pregnant uterus and so is thought to initiate and maintain labor. It also causes the glandular cells of the breast to release milk into ducts from which a baby can ob-tain it by suckling. Physicians sometimes

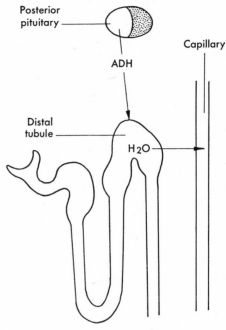

Fig. 10-3

The posterior pituitary gland secretes the anti-diuretic hormone (ADH), which accelerates kid-ney tubule reabsorption of water.

prescribe oxytocin injections to induce or to increase labor.

THE THYROID GLAND

Do you recall that earlier in this chapter we mentioned that two endocrine glands lie outside any body cavity? The thyroid is one of these. It lies in the neck just below the larynx.

Thyroid hormone influences every one of the billions of cells of our bodies. It makes them speed up their release of energy from foods. Thyroid hormone stimulates catab-olism, in other words. This has far-reaching effects. Every single function of the body depends upon a normal supply of energy, and therefore depends upon normal thy-roid secretion. Even normal mental and physical growth and development depend upon normal thyroid functioning.

If the thyroid secretes too little hormone,

catabolism slows down. As a result cells have too little energy and cannot do their work properly. If, on the other hand, the thyroid secretes too much hormone, catabolism speeds up too much. This produces very noticeable effects. The hyperthyroid individual appears nervous and too active—"hyperactive," we say.

THE PARATHYROID GLAND

The parathyroids are small glands. There are usually four of them, and they are found on the back of the thyroid. Parathyroid hormone influences the chemistry of the blood. To be specific, it causes calcium to leave the bones and enter the blood; it therefore tends to increase blood calcium. This is a matter of life and death importance because our cells are extremely sensitive to changing amounts of blood calcium. They cannot function normally either with a little too much or with a little too little calcium. For example, with too much blood calcium brain cells and heart cells soon do not function normally; a person becomes mentally disturbed and his heart may even stop. On the other hand, with too little blood calcium nerve cells become overactive—sometimes to such an extreme degree that they bombard muscles with so many impulses that the muscles go into spasms. This is called *tetany*.

THE ADRENAL GLANDS

As you can see in Figure 10-1, an adrenal gland lies curved over the top of each kidney. Although from the surface an adrenal gland appears to be only one organ, it actually is two separate endocrine glands, namely, the adrenal cortex and the adrenal medulla. Does this two-glands-in-one structure remind you of another endocrine organ? (See page 129.) The adrenal cortex is the outer part of an adrenal gland and the medulla is its inner part. Adrenal cortex hormones have different names and quite different actions from adrenal medulla hormones.

Functions of adrenal cortex hormones

Hormones secreted by the adrenal cortex are called *corticoids*. There are three kinds of corticoids: glucocorticoids (or G-C's, for short), mineralocorticoids (or M-C's), and androgens.

Glucocorticoids—chiefly the hormone called cortisol or hydrocortisone—resemble thyroid hormone in that they act on all body cells. Glucocorticoids help control some of the most vital cellular functions. Their two most important general functions are these: glucocorticoids enable cells to carry on normal metabolism of all three kinds of foods; glucocorticoids also enable cells to respond to conditions of stress in ways that make the body able to resist stress. The following paragraphs explain these two general functions of glucocorticoids a little more fully.

Glucocorticoids in normal amounts act on cells in some way to make them able to carry on normal anabolism and catabolism of proteins, fats, and carbohydrates. Larger than usual amounts of glucocorticoids accelerate both protein and fat catabolism. High blood concentrations of glucocorticoids especially accelerate the breakdown of tissue proteins to amino acids. These amino acids then move out of the cells into the blood and circulate to the liver. Liver cells change them to glucose by the process of gluconeogenesis. When this glucose newly made by liver cells enters the blood, it tends to increase the blood sugar concentration above its normal level. Excess glucocorticoids, therefore, tend to produce hyperglycemia. Excess glucocorticoids also tend to cause tissue-wasting—because they accelerate protein and fat catabolism. In fact glucocorticoids are often called "catabolic hormones." Do you recall the name of an anterior pituitary hormone that in excessive amounts tends to produce hyperglycemia?

When the body is subjected to stress—either physical or mental—the normal adrenal cortex soon starts to increase its secre-

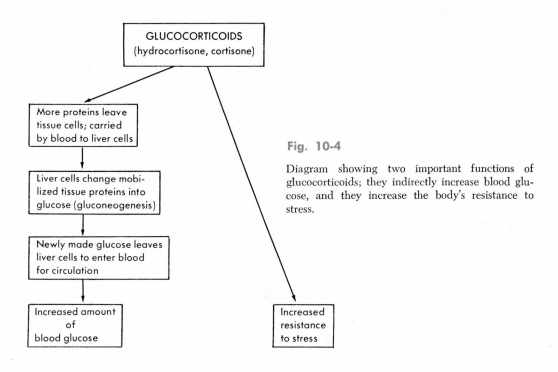

Fig. 10-4

Diagram showing two important functions of glucocorticoids; they indirectly increase blood glucose, and they increase the body's resistance to stress.

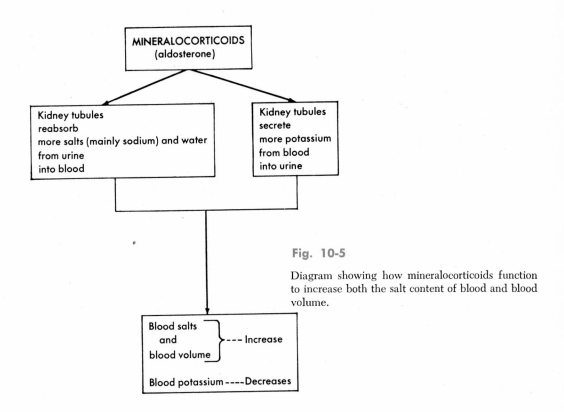

Fig. 10-5

Diagram showing how mineralocorticoids function to increase both the salt content of blood and blood volume.

tion of glucocorticoids. The high blood concentration of glucocorticoids either directly or indirectly brings about several changes known as "stress responses." Among the most important of these are the following:

1. Increased protein and fat catabolism, and after a period of time, tissue-wasting and loss of body weight;
2. Increased gluconeogenesis and high blood sugar (hyperglycemia);
3. Atrophy (decrease in size) of the thymus gland and the lymph nodes;
4. A marked decrease in the number of lymphocytes and eosinophils in the blood;
5. Decreased formation of antibodies and therefore less immunity to infectious diseases;
6. Slower formation of scar tissue and therefore slower wound healing.

Mineralocorticoids, as their name suggests, help control the amount of certain mineral salts in the blood. Aldosterone is the name of the chief M-C. Its main effects are to increase the amount of blood sodium and to decrease the amount of blood potassium. Aldosterone produces these effects by influencing the kidney tubules to speed up their reabsorption of sodium back into the blood so that less of it will be lost in the urine. At the same time, the tubules increase their secretion of potassium so that more of this mineral will be lost in the urine. Aldosterone also tends to speed up kidney reabsorption of water.

The third kind of hormone secreted by the adrenal cortex acts like the male sex hormone, testosterone. Compounds of this type, therefore, are called *androgens* (from the Greek word *andros,* meaning man). Androgens are male sex hormones; they produce masculinizing effects. Strangely enough, even a woman's adrenal glands secrete some androgens. The adrenal cortex also secretes female sex hormones under certain circumstances, but usually in such small amounts that they do not have a noticeable effect on the body.

Functions of adrenal medulla hormones

Our bodies have many ways to defend themselves against enemies that threaten their well-being. For example, the adrenal medulla responds very rapidly to many different kinds of injurious factors by greatly increasing its secretion of hormones. *Epinephrine* and *norepinephrine* are the hormones secreted by the adrenal *medulla.* They help the body meet stressful situations but are not essential for maintaining life. Glucocorticoids, the hormones from the adrenal cortex, on the other hand, both help the body resist stress and are essential for life.

Suppose that you suddenly faced some threatening situation. Imagine that you found a lump in your breast, or that your doctor told you that you had to have a dangerous operation, or that a gunman threatened to kill you. Almost instantaneously, the medullas of your two adrenal glands would be galvanized into feverish activity. They would quickly secrete large amounts of epinephrine (Adrenalin) into your blood. Many of your body functions would seem to become supercharged. Your heart would beat faster, your blood pressure would rise, more blood would be pumped to your skeletal muscles, your blood would contain more sugar for more energy, and so on. In short, you would be geared for strenuous activity, for "fight or flight," to use the words of a famous physiologist, Walter Cannon. Epinephrine prolongs and intensifies changes in body function brought about by one division of the nervous system. (Do you recall which one? Check your answer on page 59.)

THE ISLANDS OF LANGERHANS

All of the endocrine glands discussed so far are big enough to be seen without a magnifying glass or microscope. The islands of Langerhans, in contrast, are too tiny to be seen without a microscope. These glands are merely little clumps of cells scattered like islands in a sea among other pancreatic cells which secrete the pancreatic digestive juice. Paul Langerhans discovered these cell "islands" in the pancreas almost one hundred years ago—hence their name, islands of Langerhans.

Two kinds of cells in the islands of Langerhans are those called alpha and beta cells. Beta cells secrete insulin and alpha cells secrete another hormone, called glucagon, which helps control carbohydrate metabolism. *Insulin* acts primarily to accelerate the movement of glucose out of the blood into cells. Therefore, insulin tends to increase glucose catabolism and to decrease glucose concentration in the blood. *Glucagon* serves as an antagonist to insulin by accelerating liver glycogenolysis (conversion of glycogen to glucose), thereby tending to increase blood glucose.

If the islands of Langerhans secrete a normal amount of insulin, a normal amount of glucose enters the cells, and a normal amount of glucose stays behind in the blood. ("Normal blood sugar" is about 80 to 120 milligrams of glucose in every 100 milliliters of blood.) If the islands of Langerhans secrete too much insulin, as they sometimes do when a person has a tumor of the pancreas, then more glucose than usual leaves the blood to enter the cells so that blood sugar decreases. If on the other hand the islands of Langerhans secrete too little insulin, as they do in *diabetes mellitus,* less glucose leaves the blood to enter the cells so that blood sugar increases—sometimes to even three or more times the normal amount.

THE THYMUS GLAND

One of the body's best-kept secrets has been the thymus gland's functions. Before 1961, about all that was known about this organ was that it was located in the mediastinal portion of the chest, that it grew in size until a child reached the age of puberty, and that afterwards over the years it gradually became smaller. Since 1961, starting with some experiments done on newborn

mice, many facts have been discovered and ideas suggested about the thymus gland's functions. It is now known that the thymus functions in some way to make a person able to develop immunity against various diseases. Physiologists today think that the thymus probably does two things that are essential for the development of immunity. First, it probably produces the original lymphocytes formed in the body before birth and continues to produce them after birth. (Lymphocytes then travel from the thymus to the lymph nodes and spleen by way of the circulation.) Secondly, the thymus probably forms a hormone essential for immunity. Apparently this hormone must be present for a short time after a baby is born, if he is ever going to be able to become immune to any disease. Scientists have postulated that the thymus hormone acts on lymphocytes, causing them to change into plasma cells, which then form the antibodies that produce immunity.

THE FEMALE SEX GLANDS

A woman's body has two kinds of sex glands. They are the graafian follicles and the corpus luteum. Do you recall in which organ both of these endocrine glands are located? Look back in Chapter 9 if you do not remember. Check there, too, to make sure that you know which hormones the follicles and the corpus luteum secrete and what functions these hormones perform.

THE MALE SEX GLANDS

Some of the cells of the testes secrete semen (the male reproductive fluid) into ducts. Other cells referred to as *interstitial cells* secrete the male hormone, testosterone, into the blood. The interstitial cells of the testes, therefore, are the male endocrine glands. Testosterone promotes "maleness," that is, it promotes the development of the male sex organs and of the male secondary sex characteristics such as the low-pitched voice, the larger, stronger muscles, the facial beard, and so on.

Outline summary—The endocrine system

THE PITUITARY GLAND

1. Size and location—about size of a pea; lies in depression (Turk's saddle) of sphenoid bone in cranial cavity
2. Parts
   a. Anterior lobe (adenohypophysis); secretes hormones that control thyroid gland, adrenal cortex, ovaries, breasts, and testes; see Figure 10-2, page 132
   b. Posterior lobe (neurohypophysis); secretes hormone (ADH) that stimulates kidney tubules to reabsorb water back into blood at faster rate; see Figure 10-3, page 133

THE THYROID GLAND

1. Location—in neck, just below larynx
2. Function—secretes thyroid hormone which stimulates catabolism

THE PARATHYROID GLAND

1. Location—behind thyroid gland
2. Function—secretes parathyroid hormone that causes calcium to leave bone and enter blood

THE ADRENAL GLANDS

1. Location—on top of kidneys
2. Parts
   a. Adrenal cortex—outer part of adrenals; secretes hormones called corticoids; glucocorticoids enable cells to carry on normal metabolism and body to resist stress; mineralocorticoids tend to increase amount of blood sodium and decrease blood potassium; adrenal cortex also secretes small amounts of androgens (male hormones) and small amounts of female hormones
   b. Adrenal medulla—interior of adrenals; secretes epinephrine and norepinephrine, hormones that help prepare the body for emergencies of various kinds

THE ISLANDS OF LANGERHANS

1. Location—microscopic clusters of cells in pancreas
2. Function—beta cells secrete insulin, hormone that promotes cells' metabolism of glucose and therefore tends to decrease blood glucose; alpha cells secrete glucagon, hormone that accelerates glycogenolysis, and therefore tends to increase blood glucose

THE FEMALE SEX GLANDS

Ovaries contain two kinds of cells that secrete hormones—cells of gaafian follicles and of corpus luteum; see Chapter 9

THE MALE SEX GLANDS

Interstitial cells of testes secrete male hormone, testosterone, which promotes development of male sex organs and of male secondary sex characteristics

Review questions—The endocrine system

1. What endocrine gland (or glands) are located in the following parts of the body?
   cranial cavity   abdominal cavity
   neck   pelvic cavity
2. What endocrine gland is known as the "master gland"? Why?
3. What endocrine gland secretes each of the following:
   ACTH   growth hormone
   aldosterone   progesterone
   insulin
4. What hormone prepares the body for strenuous activity—for "fight or flight," in other words? *EPINEPHRINE (ADRENALIN)*
5. Many changes occur in the body when it is in a condition of stress; for example, after a person has had major surgery. Name two endocrine glands that greatly increase their secretion of hormones in times of stress. Name the hormones which they secrete.
6. Metabolism changes when the body is in a condition of stress. How does the metabolism of proteins, of fats, and of carbohydrates change and what hormones cause the changes?
7. What hormone, if a high concentration of it is present in the blood, tends to make us less immune to infectious diseases?
8. What hormone is called the "water-retaining hormone" because it decreases the amount of urine formed?
9. What hormone, if a high concentration of it is present in the blood, tends to make wounds heal more slowly?
10. What hormone is called the "salt-retaining hormone" because it makes the kidneys reabsorb sodium into the blood more rapidly, so that less sodium is lost in the urine?

11. Name two hormones or more that tend to increase blood sugar.
12. What hormone speeds up the rate of catabolism—that is, makes you burn up your foods faster?

13. What is the main function of each of the following:

    ACTH   parathyroid hormone
    ADH   thyroid hormone
    epinephrine   testosterone
    insulin

14. Which hormone is called the ovulating hormone?

Suggested supplementary readings

Anthony, Catherine P.: Textbook of anatomy and physiology, ed. 7, St. Louis, 1967, The C. V. Mosby Co.

Anthony, Catherine P.: Basic concepts in anatomy and physiology, St. Louis, 1966, The C. V. Mosby Co.

Francis, Carl C: Introduction to human anatomy, St. Louis, 1964, The C. V. Mosby Co.

Gray, Henry: Anatomy of the human body, ed. 26 (edited by Goss, Charles M.), Philadelphia, 1954, Lea & Febiger.

Guyton, Arthur C.: Textbook of medical physiology, ed. 3, Philadelphia, 1966, W. B. Saunders Co.

Ham, Arthur W.: Histology, ed. 5, Philadelphia, 1965, J. P. Lippincott Co.

Tuttle, W. W., and Schottelius, Byron A.: Textbook of physiology, ed. 15, St. Louis, 1965, The C. V. Mosby Co.

Glossary

abdomen (ab-do'men) body area between the diaphragm and pelvis.

abduct (ab-dukt') to move away from the midline; opposite of adduct.

absorption (ab-sorp'shen) passage of a substance through a membrane (for example, skin or mucosa) into blood.

acetabulum (as'e-tab'yoo-lem) socket in the hip bone (os coxa or innominate bone) into which the head of the femur fits.

Achilles tendon (A-kil'ez ten'don) tendon inserted on calcaneus; so-called because of the Greek myth that Achilles' mother held him by the heels when she dipped him in the river Styx, thereby making him invulnerable except in this area.

acidosis (as'i-do'sis) condition in which there is an excessive proportion of acid in the blood.

acromion (a-kro'mi-on) bony projection of the scapula; forms point of the shoulder.

adduct (a-dukt') to move toward the midline; opposite of abduct.

adenohypophysis (ad'e-no-hi'pof'i-sis) anterior pituitary gland.

adenoid (ad'n-oid') literally, glandlike; adenoids or pharyngeal tonsils are paired lymphoid structures in the nasopharynx.

adolescence (ad'l-es'-ns) period between puberty and adulthood.

adrenergic fibers (ad-re'ner'jic fi'bers) axons whose terminals release norepinephrine and epinephrine.

afferent neuron (af'er-ent noor'on) transmitting impulses to the central nervous system.

albuminuria (al-bu'mi-nyoor'i-a) albumin in the urine.

alkalosis (al'ka-lo'sis) condition in which there is an excessive proportion of alkali in the blood; opposite of acidosis.

alveolus (al-ve'a-les) literally a small cavity; alveoli of lungs are microscopic saclike dilatations of terminal bronchioles.

amenorrhea (a-men'o-re'a) absence of the menses.

amino acid (a-mi'no as'id) organic compound having an NH_3 and a COOH group in its molecule; has both acid and basic properties; amino acids are the structural units from which proteins are built.

amphiarthrosis (am'fi-ar-thro'sis) slightly movable joint.

141

anabolism (a-nab'o-liz'm) synthesis by cells of complex compounds (for example, protoplasm and hormones) from simpler compounds (amino acids, simple sugars, fats, and minerals); opposite of catabolism, the other phase of metabolism.

anastomosis (a-nas'to-mo'sis) connection between vessels; the circle of Willis, for example, is an anastomosis of certain cerebral arteries.

anemia (a-ne'mi-a) deficient number of red blood cells or deficient hemoglobin.

anesthesia (an'es-the'zha) loss of sensation.

aneurysm (an'yoor-iz'm) blood-filled saclike dilatation of the wall of an artery.

angina (an'ji-na) any disease characterized by spasmodic suffocative attacks; for example, angina pectoris and paroxysmal thoracic pain with feeling of suffocation.

ankylosis (an'ke-lo'sis) abnormal immobility of a joint.

anorexia (anorex'ia) loss of appetite.

anoxia (an-ok'si-a) deficient oxygen supply to tissues.

antagonistic muscles (an-tag'e-nis'tik mus'l) those having opposing actions; for example, muscles that flex the upper arm are antagonists to muscles that extend it.

anterior (an-ter'i-er) front or ventral; opposite of posterior or dorsal.

antibody, immune body (an'ti-bod'i, i-mun' bod'i) substance produced by the body that destroys or inactivates a specific substance (antigen) that has entered the body; for example, diphtheria antitoxin is the antibody against diphtheria toxin.

antigen (an'te-jen) substance which, when introduced into the body, causes formation of antibodies against it.

antiseptic (an'ti-sep'tik) preventing bacterial growth and multiplication.

antrum (an'trem) cavity; for example, the antrum of Highmore, the space in each maxillary bone, or the maxillary sinus.

anus (a'nes) distal end or outlet of the rectum.

apex (a'peks) pointed end of a conical struture.

aphasia (e-fa'zhe) loss of a language faculty such as the ability to use words or to understand them.

apnea (ap-ne'a) temporary cessation of breathing.

aponeurosis (ap'e-nyoo-ro'sis) flat sheet of white fibrous tissue that serves as a muscle attachment.

arachnoid (a-rak'noid) delicate, weblike middle membrane of the meninges.

areola (a-re'e-le) small space; the pigmented ring around the nipple.

arteriole (ar-ter'i-ol) small branch of an artery.

artery (ar'ter-i) vessel carrying blood away from the heart.

arthrosis (ar-thro'sis) joint or articulation.

articulation (ar-tik'yoo-la'shen) joint.

ascites (a-si'tez) accumulation of serous fluid in the abdominal cavity.

asphyxia (as-fik'si-a) loss of consciousness due to deficient oxygen supply.

aspirate (as'pe-rat') to remove by suction.

ataxia (a-tak'si-a) loss of power of muscle coordination.

atrium (a'tri-em) chamber or cavity; for example, atrium of each side of the heart.

atrophy (at're-fi) wasting away of tissue; decrease in size of a part.

auricle (o'ri-k'l) part of the ear attached to the side of the head; earlike appendage of each atrium of heart.

autonomic (o'te-nom'ik) self-governing; independent.

axilla (ak-sil'a) armpit.

axon (ak'son) nerve cell process that transmits impulses away from the cell body.

Bartholin (Bar'to-lin) seventeenth century Danish anatomist.

basophil (ba'-so-fil) white blood cell that stains readily, with basic dyes.

biceps (bi'seps) two headed.

bilirubin (bil'e-ru'bin) red pigment in the bile.

biliverdin (bil'e-ver'din) green pigment in the bile.

Bowman (Bo'men) nineteenth century English physician.

brachial (bra'ki-el) pertaining to the arm.

bronchiectasis (bron-ki-ek'ta-sis) dilatation of the bronchi.

bronchiole (bron'ki-ol') small branch of a bronchus.

bronchus (bron'kes) one of the two branches of the trachea.

buccal (buk'l) pertaining to the cheek.

buffer (buf'er) compound which combines with an acid or with a base to form a weaker acid or base, thereby lessening the change in hydrogen-ion concentration that would occur without the buffer.

bursa (bur'sa) fluid-containing sac or pouch lined with synovial membrane.

buttock (but'ek) prominence over the gluteal muscles.

calculus (kal'kyoo-les) stone; formed in various parts of the body; may consist of different substances.

calorie (kal′e-ri) heat unit; a large calorie is the amount of heat needed to raise the temperature of 1 kilogram of water 1 degree centigrade.

calyx (ka′liks) cup-shaped division of the renal pelvis.

capillary (kap′l-er′i) microscopic blood vessel; capillaries connect arterioles with venules; also, microscopic lymphatic vessels.

carbaminohemoglobin, carbohemoglobin (kar-ba-mi′no-he′-mo-glo′bin) compound formed by union of carbon dioxide with hemoglobin.

carbohydrate (kar′be-hi′drat) organic compounds containing carbon, hydrogen, and oxygen in certain specific proportions; for example, sugars, starches, and cellulose.

carboxyhemoglobin (kar-bok′si-he′-mo-glo′bin) compound formed by union of carbon monoxide with hemoglobin.

carcinoma (kar′se-no′ma) cancer, a malignant tumor.

caries (kar′ez) decay of teeth or of bone.

carotid (ka-rot′id) from Greek word meaning to plunge into deep sleep; carotid arteries of the neck so called because pressure on them may produce unconsciousness.

carpal (kar′p′l) pertaining to the wrist.

casein (ka′si-in) protein in milk.

cast (kast) mold; for example, formed in renal tubules.

castration (kas-tra′shen) removal of testes or ovaries.

catabolism (ka-tab′e-liz′m) breakdown of food compounds or of protoplasm into simpler compounds; opposite of anabolism, the other phase of metabolism.

catalyst (kat′l-ist) substance which accelerates the rate of a chemical reaction.

cataract (kat′a-rakt′) opacity of the lens of the eye.

catecholamines (kat-e-ko-la′mins) norepinephrine and epinephrine.

cecum (se′kum) blind pouch; the pouch at the proximal end of the large intestine.

celiac (se′le-ak) pertaining to the abdomen.

cellulose (sel′u-los) polysaccharide, the main plant carbohydrate.

centimeter (sen′ti-me′ter) 1/100 of a meter, about 2/5 of an inch.

centrioles (sen′tri-ols) two dots seen with a light microscope in the centrosphere of a cell; active during mitosis.

cerumen (se-ru′men) earwax.

cervix (ser′viks) neck; any neckline structure.

chiasm (ki′azm) crossing; specifically, a crossing of the optic nerves.

cholecystectomy (ko-le-sis-tek′to-me) removal of the gallbladder.

cholesterol (ko-les′te-rol′) organic alcohol present in bile, blood, and various tissues.

cholinergic fibers (ko-li′ner-jic fi′bers) axons whose terminals release acetylcholine.

cholinesterase (ko-li′nes′ter-ase) enzyme; catalyzes breakdown of acetylcholine.

chromatin (kro′ma-tin) deep-staining substance in the nucleus of cells; divides into chromosomes during mitosis.

chromosome (kro′mo-som) one of the segments into which chromatin divides during mitosis; involved in transmitting hereditary characteristics.

chyle (kil) milky fluid; the fat-containing lymph in the lymphatics of the intestine.

chyme (kim) partially digested food mixture leaving the stomach.

cilia (sil′e-ah) hairlike projections of protoplasm.

circadian (ser′ka-de′an) daily.

cochlea (kok′le-ah) snail shell or structure of similar shape.

coenzyme (ko-en′zime) nonprotein substance which activates an enzyme.

collagen (kol′la-jen) principle organic constituent of connective tissue.

colloid (kol′oid) dissolved particles with diameters of 1 to 100 millimicrons (1 millimicron equals about 1/25,000,000 of an inch).

colostrum (ko-los′trum) first milk secreted after childbirth.

concha (kong′kah) shell-shaped structure; for example, bony projections into the nasal cavity.

condyle (kon′dil) rounded projection at the end of a bone.

congenital (kon-jen′it-al) present at birth.

contralateral (kon′tra-lat′er-el) on the opposite side.

coracoid (kor′ak-oid) like a raven's beak in form.

corium (kor′e-um) true skin or derma.

coronal (kor′e-n′l) like a crown.

coronary (kor′e-ner′i) encircling; in the form of a crown.

corpus (kor′pes) body.

corpuscle (kor′pus′l) very small body or particle.

cortex (kor′teks) outer part of an internal organ; for example, of the cerebrum and of the kidneys.

cortisol (kor′ti-sol) the chief hormone secreted

by the adrenal cortex; hydrocortisone; compound F.

costal (kos'tal) pertaining to the ribs.

crenation, plasmolysis (kre-na'shun, plaz-mol' is-is) shriveling of a cell due to water withdrawal.

cretinism (kre'tin-iz'm) dwarfism due to hypofunction of the thyroid gland.

cribriform (krib'rif-orm) sievelike.

cricoid (kri'koid) ring-shaped; a cartilage of this shape in the larynx.

crystalloid (kris't'l-oid') disolved particle less than 1 millimicron in diameter.

cutaneous (ku-ta'ne-us) pertaining to the skin.

cyanosis (si-an-o'sis) bluish appearance of the skin due to deficient oxygenation of the blood.

cytology (si-tol'o-ji) study of cells.

cytoplasm (si'to-plazm) the protoplasm of a cell exclusive of the nucleus.

deciduous (de-sid'u-us) temporary; shedding at a certain stage of growth; for example, deciduous teeth.

decussation (de-cus-sa'shun) crossing over like an X.

defecation (def-ek-a'shun) elimination of waste matter from the intestines.

deglutition (deg-lu-tish'un) swallowing.

deltoid (del'toid) triangular; for example, deltoid muscle.

dendrite, dendron (den'drit, den'dron) branching or treelike; a nerve cell process that transmits impulses toward the cell body.

dens (denz) tooth.

dentate (den'tat) having toothlike projections.

dentine (den'ten) main part of a tooth, under the enamel.

dentition (den-tish'en) teething; also, number, shape, and arrangement of the teeth.

dermis, corium (der'mis, kor'e-um) true skin.

dextrose (deks'tros) glucose, a monosaccharide, the principal blood sugar.

diaphragm (di'a-fram) membrane or partition that separates one thing from another; the muscular partition between the thorax and abdomen; the midriff.

diaphysis (di-af'i-sis) shaft of a long bone.

diarthrosis (di'ar-thro'sis) freely movable joint.

diastole (di-as'to-le) relaxation of the heart, interposed between its contractions; opposite of systole.

diencephalon (di-en-sef'a-lon) "tween" brain; parts of the brain between the cerebral hemispheres and the mesencephalon or midbrain.

diffusion (di-fu'zhen) spreading; for example, scattering of dissolved particles.

digestion (di-jes'chun) conversion of food into assimilable compounds.

diplopia (dip-lo'pe-ah) double vision; seeing one object as two.

disaccharide (di-sak'e-rid') sugar formed by the union of two monosaccharids; contains twelve carbon atoms.

distal (dis'tal) toward the end of a structure; opposite of proximal.

dorsal, posterior (dor'sal, pos-te're-or) pertainting to the back; opposite of ventral.

dropsy (drop'si) accumulation of serous fluid in a body cavity or in tissues; edema.

dura mater (dyoor'a ma'ter) literally strong or hard mother; outermost layer of the meninges.

dyspnea (disp-ne'ah) difficult or labored breathing.

dystrophy (dis-tro'fi) faulty nutrition.

ectopic (ek-top'ik) displaced; not in the normal place; for example, extrauterine pregnancy.

edema (e-de'mah) excessive fluid in tissues; dropsy.

effector (e-fek'ter) responding organ; for example, voluntary and involuntary muscle, the heart, and glands.

efferent (ef'er-ent) carrying from, as neurons which transmit impulses from the central nervous system to the periphery; opposite of afferent.

electrocardiogram (e-lek'tro-kar'di-o-gram) graphic record of heart's action potentials.

electroencephalogram (e-lek'tro-en-cef'a-logram) graphic record of brain's action potentials.

electrolyte (i-lek'tre-lit') substance that ionizes in solution, rendering the solution capable of conducting an electric current.

electron (e-lek'tron) minute, negatively charged particle.

elimination (i-lim'a-na'shen) expulsion of wastes from the body.

embolism (em'be-liz'm) obstruction of a blood vessel by foreign matter carried in the bloodstream.

embryo (em'bre-o) animal in early stages of intrauterine development; the human fetus the first three months after conception.

emesis (em'e-sis) vomiting.

emphysema (em'fi-se'ma) dilatation of pulmonary alveoli.

empyema (em-pi-e'mah) pus in a cavity; for example, in the chest cavity.

encephalon (en-sef'a-lon') brain.

endocrine (en'da-krin') secreting into the blood or tissue fluid rather than into a duct; opposite of exocrine.

energy (en'er-ji) capacity for doing work.

enteron (en'te-ron) intestine.

enzyme (en'zim) catalytic agent formed in living cells.

eosinophil, acidophil (e'o-sin-o-fil, a-sid'o-fil) white blood cell readily stained by eosin.

epidermis (ep'e-dur'mis) "false" skin; outermost layer of the skin.

epinephrine (ep-e-nef'rin) adrenalin; secretion of the adrenal medulla.

epiphyses (e-pif'i-ses) ends of a long bone.

erythrocyte (e-rith'ro-sit) red blood cell.

ethmoid (eth'moid) sievelike.

eupnea (up-ne'ah) normal respiration.

Eustachio (yoo-sta'shi-o) Italian anatomist of the sixteenth century.

exocrine (eks-o'krin) secreting into a duct; opposite of endocrine.

exophthalmos (eks-of-thal'mus) abnormal protrusion of the eyes.

extrinsic (eks-trin'sik) coming from the outside; opposite of intrinsic.

Fallopius (Fal-lo'pius) sixteenth century Italian anatomist.

fascina (fash'e-ah) sheet of connective tissue.

fasciculus (fa-sik'u-lus) little bundle.

fetus (fe'tes) unborn young, especially in the later stages; in human beings, from third month of intrauterine period until birth.

fiber (fi'-ber) threadlike structure.

fibrin (fi'brin) insoluble protein in clotted blood.

fibrinogen (fi'rin'e-jen) soluble blood protein which is converted to insoluble fibrin during clotting.

fimbria (fim'bri-a) fringe.

fissure (fish'er) groove.

flaccid (flak'sid) soft, limp.

folicle (fol'i-k') small sac or gland.

fontanelle (fon'te-nel') "soft spots" of the infant's head; unossified areas in the infant skull.

foramen (fo-ra'men) small opening.

fossa (fos'a) cavity or hollow.

fovea (fo'vi-a) small pit or depression.

fundus (fun'des) base of a hollow organ; for example, the part farthest from its outlet.

ganglion (gang'gli-on) cluster of nerve cell bodies outside the central nervous system.

gasserian (gas-se'ri-an) named for Gasser, a sixteenth century Austrian surgeon.

gastric (gas'trik) pertaining to the stomach.

gene (jen) part of the chromosome that transmits a given hereditary trait.

genitals (jen'a-t'ls) reproductive organs; genitalia.

gestation (jes-ta'shen) pregnancy.

gland (gland) secreting structure.

glomerulus (glo-mer'u-lus) compact cluster; for example, of capillaries in the kidneys.

glossal (glos'l) of the tongue.

glucagon (glu-ka-gon) hormone secreted by alpha cells of the islands of Langerhans.

glucocorticoids (glu-ko-kor'ti-koids) hormones that influence food metabolism; secreted by adrenal cortex.

gluconeogenesis (glu'ko-ne-o-jen'e-sis) formation of glucose from protein or fat compounds.

glucose (gloo'kos) monosccaccharide or simple sugar; the principal blood sugar.

gluteal (gloo-te'el) of or near the buttocks.

glycerin, glycerol (glis'er-in, glis'er-ol) product of fat digestion.

glycogen (gli'ke-jen) polysaccharide; animal starch.

glycogenesis (gli'ko-jen'e-sis) formation of glycogen from glucose or from other monosaccharides, fructose or galactose.

glycogenolysis (gli'ko-je-nol'i-sis) hydrolysis of glycogen to glucose-6-phosphate or to glucose.

glyconeogenesis (gli'ko-ne-o-jen'e-sis) the formation of glycogen from protein or fat compounds.

gonad (gon'ad) sex gland in which reproductive cells are formed.

graafian (graaf-ian) named for Graaf, a seventeenth century Dutch anatomist.

gradient (gra'di-ent) a slope or difference between two levels; for example, blood pressure gradient—a difference between the blood pressure in two different vessels.

gustatory (gus'te-tor'i) pertaining to taste.

gyrus (ji'rus) convoluted ridge.

haversian (ha-ver'shan) named for Havers, English anatomist of the late seventeenth century.

hemiplegia (hem-e-ple'je-ah) paralysis of one side of the body.

hemoglobin (he'ma-glo'bin) iron-containing protein in red blood cells.

hemolysis (he-mol'i-sis) destruction of red blood cells with escape of hemoglobin from them into surrounding medium.

hemopoiesis (he'mo-poy-e'sis) blood cell formation.

hemorrhage (hem'o-rej) bleeding.

hepar (he'par) liver.

heparin (hep'er-in) substances obtained from the liver which inhibits blood clotting.

heredity (he-red'a-ti) transmission of characteristics from a parent to a child.

hernia, "rupture" (hur'ni-a, "rup'cher") protrusion of a loop of an organ through an abnormal opening.

hilus, hilum (hi'lus, hi'lem) depression where vessels enter an organ.

His (His) German anatomist of the late nineteenth century.

histology (his-tol'e-ji) science of minute structure of tissues.

homeostasis (ho-me-os'tas-is) relative uniformity of the normal body's internal environment.

hormone (hor'mon) substance secreted by an endocrine gland.

hyaline (hi'a-lin) glasslike.

hydrocortisone (hi'dro-kor'ti-son) a hormone secreted by the adrenal cortex; cortisol; compound F.

hydrolysis (hi-drol'a-sis) literally "split by water"; chemical reaction in which a compound reacts with water to form simpler compounds.

hymen (hi'men) Greek for skin; mucous membrane that may partially or entirely occlude the vaginal outlet.

hyoid (hi'oid) U-shaped; bone of this shape at the base of the tongue.

hyperemia (hi'per-e'mi-a) increased blood in a part.

hyperkalemia (hi'per-ka-le'mi-a) higher than normal concentration of potassium in the blood.

hypernatremia (hi'per-na-tre-mi-a) higher than normal concentration of sodium in the blood.

hyperopia (hi-per-o'pi-a) farsightedness.

hyperplasia (hi-per-pla'zhi-a) increase in the size of a part due to an increase in the number of its cells.

hyperpnea (hi'perp-ne'a) abnormally rapid breathing; panting.

hypertension (hi'per-ten'shen) abnormally high blood pressure.

hyperthermia (hi-per-ther'mi-a) fever; body temperature above 37° C.

hypertrophy (hi-pur'tro-fi) increased size of a part due to an increase in the size of its cells.

hypokalemia (hi'po-ka-le'mi-a) lower than normal concentration of potassium in the blood.

hyponatremia (hi'po-na-tre'mi-a) lower than normal concentration of sodium in the blood.

hypophysis (hi-pof'a-sis) Greek for undergrowth; hence the pituitary gland which grows out from the undersurface of brain.

hypothermia (hi'po-ther'mi-a) subnormal body temperature below 37° C.

hypoxia (hi'pok'si-a) oxygen deficiency.

incus (in'kes) anvil; the middle ear bone which is shaped like an anvil.

inferior (in-fer'i-er) lower; opposite of superior.

inguinal (in'gwi-n'l) of the groin.

inhalation (in'ha-la'shen) inspiration or breathing in; opposite of exhalation or expiration.

inhibition (in'hi-bish'en) checking or restraining of action.

innominate (i-nom'a-nit) not named, anonymous; for example, ossa coxae (hip bones) formerly known as innominate bones.

insulin (in'sa-lin) hormone secreted by islands of Langerhans in the pancreas.

intercellular (in'ter-sel'yoo-ler) between cells; interstitial.

internuncial (in'ter-nun'si-al) like a messenger between two parties; hence an internuncial neuron (or interneuron) is one which conducts impulses from one neuron to another.

interstitial (in-tur-stish'al) forming small spaces between things; intercellular.

intrinsic (in-trin'sik) not dependent upon externals; located within something; opposite of extrinsic.

involuntary (in-vol'en-ter'i) not willed; opposite of voluntary.

involution (in'vo-loo'shen) return of an organ to its normal size after enlargement; also retrograde or degenerative change.

ion (i'en) electrically charged atom or group of atoms.

ipsilateral (ip-si-lat'er-el) on the same side; opposite of contralateral.

irritability (ir'e-te-bil'e-ti) excitability; ability to react to a stimulus.

ischemia (is-ke'mi-a) local anemia; temporary lack of blood supply to an area.

isotonic (i'se-ton'ik) of the same tension or pressure.

ketones (ke'tons) acids (acetoacetic, beta-hydroxybutyric, and acetone) produced during fat catabolism.

ketosis (ke-to'sis) excess amount of ketone bodies in the blood.

kilogram (kil'a-gram) 1,000 grams; approximately 2.2 pounds.

kinesthesia (kin'is-the'zha) "muscle sense";

that is, sense of position and movement of body parts.

labia (la'bi-a) lips.

lacrimal (lak're-m'l) pertaining to tears.

lactation (lak-ta'shen) secretion of milk.

lactose (lak'tos) milk sugar, a disaccharide.

lacuna (le-ku'na) space or cavity; for example, lacunae in bone contain bone cells.

lamella (le-mel'a) thin layer, as of bone.

lateral (lat'er-el) of or toward the side; opposite of medial.

leukocyte (loo'ke-sit') white blood cell.

ligament (lig'a-ment) bond or band connecting two objects; in anatomy a band of white fibrous tissue connecting bones.

lipid (li'pid) fats and fatlike compounds.

loin (loin) part of the back between the ribs and hip bones.

lumbar (lum'ber) of or near the loins.

lumen (loo'men) passageway or space within a tubular structure.

luteum (lu'te-um) golden yellow.

lymph (limf) watery fluid in the lymphatic vessels.

lymphocyte (lim'fe-sit') one type of white blood cells.

lysosomes (li'so-soms) membranous organelles containing various enzymes that can dissolve most cellular compounds; hence called "digestive bags" or "suicide bags" of cells.

malleolus (ma-le'a-les) small hammer; projections at the distal ends of the tibia and fibula.

malleus (mal'i-es) hammer; the tiny middle ear bone that is shaped like a hammer.

Malpighi (Mahl-pe'ge) seventeenth century Italian anatomist.

maltose (mawl'tos) disaccharide or "double" sugar.

mammary (mam'er-e) pertaining to the breast.

manometer (ma-nom'e-ter) instrument used for measuring the pressure of fluids.

manubrium (ma-noo'bri-em) handle; upper part of the sternum.

mastication (mas'te-ka'shen) chewing.

matrix (ma'triks) ground substance in which cells are embedded.

meatus (mi-a'tes) passageway.

medial (me'di-el) of or toward the middle; opposite of lateral.

mediastinum (me'di-as-ti'nem) middle section of the thorax; that is, between the two lungs.

medulla (mi-dul'a) Latin for marrow; hence the inner portion of an organ in contrast to the outer portion or cortex.

meiosis (mi-o'sis) nuclear division in which the number of chromosomes are reduced to half their original number before the cell divides in two.

membrane (mem'bran) thin layer or sheet.

menstruation (men'stroo-a'shen) monthly discharge of blood from the uterus.

mesentery (mes'n-ter'i) fold of peritoneum that attaches the intestine to the posterior abdominal wall.

mesial (me'zi-el) situated in the middle; median.

metabolism (me-tab'e-liz'm) complex process by which food is utilized by a living organism.

metacarpus (met'e-kar'pes) "after" the wrist; hence, the part of the hand between the wrist and fingers.

metatarsus (met'e-tar'ses) "after" the instep; hence, the part of the foot between the tarsal bones and toes.

meter (me'ter) about 39.5 inches.

microglia (mi-krog'li-a) one type of connective tissue found in the brain and cord.

micron (mi'kron) 1/1,000 millimeter; about 1/25,000 inch.

micturition (mik'tu-rish'en) urination, voiding.

millimeter (mil'li-me-ter) 1/1,000 meter; about 1/25 inch.

mineralocorticoids (min'er-al'o-kor'ti-koids) hormones that influence mineral salt metabolism; secreted by adrenal cortex; aldosterone is the chief mineralocorticoid.

mitochondria (mit'o-kon'dri-a) threadlike structures.

motoneurons (mo-to-noo'rons) transmit nerve impulses away from the brain or spinal cord; also called motor or efferent neurons.

mitosis (mi-to'sis) indirect cell division involving complex changes in the nucleus.

mitral (mi'trel) shaped like a miter.

monosaccharide (mon'e-sak'e-rid') simple sugar; for example, glucose.

myelin (mi'e-lin) lipoid substance found in the myelin sheath around some nerve fibers.

myocardium (mi'o-kar'di-em) muscle of the heart.

myopia (mi-o'pi-a) nearsightedness.

nares (nar'ez) nostrils.

neurilemma (noor'i-lem'a) nerve sheath.

neurohypophysis (noo'ro-hi-pof'i-sis) posterior pituitary gland.

neuron (noor'on) nerve cell, including its processes.

neutrophil (nu'tro-fil) white blood cell that stains readily with neutral dyes.

nucleus (noo'kli-es) spherical structure within a cell; a group of neuron cell bodies in the brain or cord.

occiput (ok'si-put) back of the head.

olecranon (o-lek're-non) elbow.

olfactory (ol-fak'ter-i) pertaining to the sense of smell

ophthalmic (of-thal'mik) pertaining to the eyes

organelle (or'gan-el) cell organ; one of the specialized parts of a single-celled organism (protozoon), serving for the performance of some individual function.

os (os) Latin for mouth and for bone.

osmosis (oz-mo'sis) movement of a fluid through a semipermeable membrane.

ossicle (os'i-k'l) little bone.

oxidation (ok'-si-da'shun) loss of hydrogen or electrons from a compound or element.

oxyhemoglobin (ok'-si-hem'-o-glo'bin) a compound formed by union of oxygen with hemoglobin.

palate (pal'it) roof of the mouth.

palpebrae (pal'pe-bra) eyelids.

papilla (pa-pil'a) small nipple-shaped elevation.

paralysis (pe-ral'e-sis) loss of the power of motion or sensation, especially loss of voluntary motion.

parenchyma (pe-ren'ki-ma) essential or functional tissue of an organ.

parietal (pa-ri'e-t'l) of the walls of an organ or cavity.

parotid (pe-rot'id) located near the ear.

parturition (par-tu-rish'un) act of giving birth to an infant.

patella (pa-tel'a) small, shallow pan; the kneecap.

pectoral (pek'to-rel) pertaining to the chest or breast.

pelvis (pel'vis) basin or funnel-shaped structure.

peripheral (pa-rif'er-el) pertaining to an outside surface.

pH (pe'ach') hydrogen-ion concentration.

phagocytosis (fag'o-si-to'sis) ingestion and digestion of particles by a cell.

phalanges (fe-lan'jez) finger or toe bones.

phrenic (fren'ik) pertaining to the diaphragm.

pia mater (pi'a ma'ter) gentle mother; the vascular innermost covering (meninges) of the brain and cord.

pineal (pin'i-el) shaped like a pine cone.

piriformis (pir-i-for'mis) pear-shaped.

pisiform (pi'sa-form') pea-shaped.

plantar (plan'ter) pertaining to the sole of the foot.

plasma (plaz'ma) liquid part of the blood.

plasmolysis (plaz-mol'a-sis) shrinking of a cell due to water loss by osmosis.

plexus (plek'ses) network.

polymorphonuclear (pol'i-mor-fo-nu'kle-ar) having many-shaped nuclei.

polysaccharide (pol-i-sak'ar-id) complex sugar.

pons (ponz) bridge.

popliteal (pop-lit'i-el) behind the knee.

posterior (pos-ter'i-er) following after; hence, located behind; opposite of anterior.

presbyopia (prez'bi-o'pi-a) "oldsightedness"; farsightedness of old age.

pronate (pro'nat) to turn palm downward.

proprioreceptors (pro'pri-o-re-sep'tors) receptors located in the muscles, tendons, and joints.

protoplasm (pro'te-plaz'm) living substance.

proximal (prok'sa-m'l) next or nearest; located nearest the center of the body or the point of attachment of a structure.

psoas (so'es) pertaining to the loin, the part of the back between the ribs and hip bones.

psychosomatic (si-ko-so-mat'ik) pertaining to the influence of the mind, notably the emotions, on body functions.

puberty (pu'ber-ti) age at which the reproductive organs become functional.

receptor (ri-sep'ter) peripheral beginning of a sensory neuron's dendrite.

reflex (re'fleks) involuntary action.

refraction (ri-frak'shen) bending of a ray of light as it passes from a medium of one density to one of a different density.

renal (re'nal) pertaining to the kidney.

reticular (ri-tik'yoo-ler) netlike.

reticulum (ri-tik'yoo-lum) a network.

ribosomes (ri'bo-soms) organelles in cyptoplasm of cells; synthesize proteins, so nicknamed "protein factories".

rugae (roo'-je) wrinkles or folds.

sagittal (saj'i-tal) like an arrow; longitudinal.

salpinx (sal'pinks) tube; oviduct.

sartorius (sar-tor'i-es) tailor; hence, the thigh muscle used to sit cross-legged like a tailor.

sciatic (si-at'ik) pertaining to the ischium.

sclera (skler'a) from Greek for hard.

scrotum (skro'tem) bag.

sebum (se'bem) Latin for tallow; secretion of sebaceous glands.

sella turcica (sel'la tur'ki'ka) Turkish saddle; saddle-shaped depression in the sphenoid bone.

semen (se'men) Latin for seed; male reproductive fluid.

semilunar (sem'i-loo'ner) half-moon shaped.

senescence (se-nes'ns) old age.

serratus (ser-ra'tus) saw-toothed.

serum (ser'em) any watery animal fluid; clear, yellowish liquid that separates from a clot of blood.

sesamoid (ses'a-moid') shaped like a sesame seed.

sigmoid (sig'moid) s-shaped.

sinus (si'nes) cavity.

soleus (so'le-us) pertaining to a sole; a muscle in the leg shaped like the sole of a shoe.

somatic (so-mat'ik) of the body framework or walls, as distinguished from the viscera or internal organs.

sphenoid sfe'noid) wedged-shaped

sphincter (sfink'ter) ring-shaped muscle.

splanchnic (splank'nik) visceral.

squamous (skwa'mes) scalelike.

stapes (sta'pez) stirrup; tiny stirrup-shaped bone in the middle ear.

stimulus stim'yoo-les) agent that causes a change in the activity of a structure.

stress (stres) according to Selye, physiological stress is a condition in the body produced by all kinds of injurious factors that he calls "stressors" and manifested by a syndrome (a group of symptoms that occur together).

stressor (stres'or) any injurious factor that produces biological stress; for example, emotional trauma, infections, severe exercise.

striated (stri'at-id) marked with parallel lines.

sudoriferous (soo'do-rif'er-es) secreting sweat.

sulcus (sul'kes) furrow or groove.

superior (se-per'i-er) higher; opposite of inferior.

supinate (soo'pa-nat') to turn the palm of the hand upward; opposite of pronate.

Sylvius (Syl'vius) seventeenth century anatomist.

symphysis (sim'fe-sis) Greek for growing together.

synapse (si-naps') joining; point of contact between adjacent neurons.

synovia (si-no'vi-a) literally "with egg"; secretion of the synovial membrane; resembles egg white.

synthesis (syn'the-sis) putting together of parts to form a more complex whole.

systole (sis'to-le') contraction of the heart muscle.

talus (ta'les) ankle; one of the bones of the ankle.

tarsus (tar'ses) instep.

tendon (ten'den) band or cord of fibrous connective tissue which attaches a muscle to a bone or other structure.

thorax (tho'raks) chest.

thrombosis (throm-bo'sis) formation of a clot in a blood vessel.

tibia (tib'i-a) Latin for shin bone.

tonus (to'nes) continued, partial contraction of muscle.

tract (trakt) bundle of axons located within the central nervous system.

trauma (trau'ma) injury.

trochlear (trok'li-er) pertaining to a pulley.

trophic (trof'ik) having to do with nutrition.

turbinate (tur'bi-nat) shaped like a cone or like a scroll or spiral.

tympanum (tim'pa-nem) drum.

umbilicus (um-bil'i-kes) navel.

utricle (u'tri-k'l) little sac.

uvula (u'vyoo-la) Latin for a little grape; a projection hanging from the soft palate.

vagina (va-ji'na) sheath.

vagues (va'ges) Latin for wandering.

valve (valv) structure which permits flow of a fluid in one direction only.

vas (vas) vessel or duct.

vastus (vas'tus) wide; of great size.

vein (van) vessel carrying blood toward the heart.

ventral (van'trel) of or near the belly; in man, front or anterior; opposite of dorsal or posterior.

ventricle (ven'tri-k'l) small cavity.

vermiform (vur'me-form') worm-shaped.

villus (vil'es) hairlike projection.

viscera (vis'er-a) internal organs.

vomer (vo'mer) ploughshare.

xiphoid (zif'oid) sword-shaped.

zygoma (zi-go'ma) yoke.

Index

Somatotrophic hormone (STH), 131-133
Spermatic artery, 75
Spermatogenesis, 119
Spermatozoa or spermatozoon, 117
Sphenoid bone, 25
Sphenoid sinuses, 25
Sphincter
  of anus, 91
  pyloric, 90
Sphincter muscle, 91
Spinal accessory nerve, 57
Spinal cavity, 3
  organs of, 4
Spinal cord, 49, 55-56
  coverings and fluid spaces of, 50
  diagram of, 51
  functions of, 55-56
  structure of, 55
  tracts, 56
Spinal nerves, 56-57
Spine, 28
  function of, 28
  sections of, 28
  structure of, 28
Spirometer, 106
Spleen, 84
Spurs, 31
Stapes, 25, 61
Steapsin, 95
Sternum, 26, 28
Stirrups; see Stapes
Stomach, 88-90
  mucous membrane of, 89
  structure of, 88-90
Stomach muscle, 89
Stress responses, 136
Striated muscle, 33, 34-42, 91
  contractions of, 35-42
  functions of, 34-35
  structure, 34-35
Subclavian artery, 75
Subclavian vein, 76
Sublingual glands, 92
Submaxillary glands, 92
Sucrase, 95
Sucrose, 95
Sugar
  in blood, 93, 137
  digestion of, 95
"Sunday throat," 102
Superficial muscle, 40-42
Suppression of urine, 113
"Swayback"; see Lordosis
Sweat glands, 17-18
  function, 18
Sympathetic nervous system, 58-59
Symphysis pubis, 26
Synapse, 48
Synergists muscles, 35
Synovial fluid, 31

Synovial membrane, 31
System
  body, organs of, 6
  definition of, 8
  types of, 8

T

Tarsal bones, 26
Teeth, 91, 92
Temporal arteries, 82
Temporal bones, 25
Tendons, 34
Testes, 117-119, 120
Testicle; see Testis
Testis, 118
Testosterone, 118, 137
Tetany, 134
Thalamus
  functions of, 54-55
  structure, 54
Thoracentesis, 5
Thoracic aorta, 75
  lining of, 6
Thoracic cavity, 3, 4, 6
  organs of, 5
Thoracic duct, 82
Thoracic vertebrae, 25
Thorax, 26, 28
Throat; see Pharynx
Thrombin, 71
Thrombokinase, 71
Thromboplastin, 71
Thrombosis, 71
Thrombus, 71
Thymus gland, 137
Thyroid cartilage, 102
Thyroid gland, 130, 133-134
Thyroid hormone, 133-134
Thyrotrophic hormone (TH), 131
Tibia, 26
Tibial arteries, 75
Tibial veins, 76
Tibialis anterior muscle, 41
Tidal air, 106
Tissue (s), 13-15
  adipose, 124
  areolar, 14
  capillaries, 104, 105
  cardiac muscle, 72
  connective, 14-15, 68
  definition of, 7
  epithelial, 13-14
  erectile, 119
  exchange of gases in, 105
  fibrous, 14
  hemopoietic, 14
  lymphatic, 68
  muscle, 15, 33-34
  myeloid, 68
Tissue fluid, 82
T.M.R., 97
Tonsils, 102
Tonus (tone), 42

Total metabolic rate, 97
Trachea, 101, 102
Tracts, 91
Transfusions, 70
Transverse colon, 91
Trapezius muscle, 38, 39
Triceps brachii, 37, 41
Tricuspid teeth, 92
Tricuspid valve, 73
Trifacial nerve, 57
Trigeminal nerve, 57
Trypsin, 94
Tuberculosis, 106
Tubules, 109
  convoluted, 110
  renal, 110
  seminiferous, 118
Turbinate bones, 25
Turk's saddle, 25, 131
Tympanic cavity, 61-62
Tympanic membrane, 61

U

Ulna, 26
Ulnar artery, 75
Ulnar vein, 76
"Universal donor" blood, 70
"Universal recipient," 70
Ureters, 113
Urethra
  female, 113, 119
  male, 113, 118, 119
Urinary bladder, 113
Urinary meatus, 113
Urinary organs, 110
Urinary retention, 113
Urinary suppression, 113
Urinary system, 109-114
  functions of, 109
  review questions, 114
  summary, 113-114
Urine, 112-113
  abnormal amounts of, 113
  formation, 112
Uterine tubes, 119, 121
Uterus, 119, 121-124
  body of, 122
  fundus of, 112

V

Vagina, 119, 124
Vaginitis, 125
Vagus nerve, 57
Vas deferens; see Seminal ducts
Vasomotor center neurons, 54
Veins, 76, 77
  coats of, 74
  function of, 77
  hepatic, 79
  venae cavae, 77
Vena cava, 76, 77
Venous blood, 79, 105
Venous plexus, 124